Constructions of Cancer in Early Modern England

Palgrave Studies in Literature, Science and Medicine

Series editors: Professor **Sharon Ruston** (Lancaster University, UK), Professor **Alice Jenkins** (University of Glasgow, UK) and Professor **Catherine Belling** (Northwestern University, USA)

Palgrave Studies in Literature, Science and Medicine is an exciting new series that focuses on one of the most vibrant and interdisciplinary areas in literary studies. Comprising academic monographs, essay collections, and Palgrave Pivot books, the series will emphasize a historical approach to its subjects. The series will cover all aspects of this rich and varied field and is open to new and emerging topics as well as established ones.

Sharon Ruston is Chair in Romanticism and Research Director for the Department of English and Creative Writing at Lancaster University, UK.

Alice Jenkins is Professor of Victorian Literature and Culture at Glasgow University, UK. She is a co-founder and former Chair of the British Society for Literature and Science.

Catherine Belling is Associate Professor in Medical Humanities and Bioethics at Northwestern University, USA. She is also the Executive Editor of the journal *Literature and Medicine*.

Editorial Board:

Steven Connor, Professor of English, University of Cambridge, UK; Lisa Diedrich, Associate Professor in Women's and Gender Studies, Stony Brook University, USA; N Katherine Hayles, Professor of English, Duke University, USA; Peter Middleton, Professor of English, University of Southampton, UK; Sally Shuttleworth, Professorial Fellow in English, St Anne's College, University of Oxford, UK; Susan Squier, Professor of Women's Studies and English, Pennsylvania State University, USA; Martin Willis, Professor of Science, Literature and Communication, University of Westminster, UK

Titles include:

Markus Iseli
THOMAS DE QUINCEY AND THE COGNITIVE UNCONSCIOUS

Esther L. Jones
MEDICINE AND ETHICS IN BLACK WOMEN'S SPECULATIVE FICTION

Ewa Barbara Luczak
BREEDING AND EUGENICS IN THE AMERICAN LITERARY IMAGINATION
Heredity Rules in the Twentieth Century

Alanna Skuse
CONSTRUCTIONS OF CANCER IN EARLY MODERN ENGLAND (Open Access)
Ravenous Natures

Palgrave Studies in Literature, Science and Medicine
Series Standing Order ISBN 978–1–1374–4538–4 hardback 978–1–1374–4543–8 paperback
(*outside North America only*)

You can receive future titles in this series as they are published by placing a standing order. Please contact your bookseller or, in case of difficulty, write to us at the address below with your name and address, the title of the series and one of the ISBNs quoted above.

Customer Services Department, Macmillan Distribution Ltd, Houndmills, Basingstoke, Hampshire RG21 6XS, England

Constructions of Cancer in Early Modern England

Ravenous Natures

Alanna Skuse
Fellow at the Folger Shakespeare Library, Washington DC

palgrave
macmillan

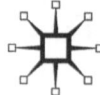

First published 2015 by
PALGRAVE MACMILLAN

Palgrave Macmillan in the UK is an imprint of Macmillan Publishers Limited, registered in England, company number 785998, of Houndmills, Basingstoke, Hampshire RG21 6XS.

Palgrave Macmillan in the US is a division of Nature America, Inc., 75 Varick Street, New York, NY 10010.

Palgrave Macmillan is the global academic imprint of the above companies and has companies and representatives throughout the world.

Palgrave® and Macmillan® are registered trademarks in the United States, the United Kingdom, Europe and other countries.

DOI 10.1057/9781137487537
E-PDF ISBN 9781137487537
E-PUB ISBN 9781137487544

Hardback ISBN 9781137487520
Paperback ISBN 9781137569196

A catalogue record for this book is available from the British Library.

A catalogue record for this book is available from the Library of Congress.

Contents

List of Illustrations

Acknowledgements

The unstinting support and wisdom of Sarah Toulalan and Andrew McRae has seen me through the PhD thesis and the wilds of early career research in which this book was constructed, and I am immeasurably grateful. Others who have offered invaluable advice and encouragement include Philip Schwyzer, Sujata Iyengar, Jennifer Evans, Sally Templeman, Margaret Healy, Karen Edwards and Lesel Dawson.

This work was supported by the Wellcome Trust [093090] and would not have been possible without their generosity. Material related to this book has been printed in *Social History of Medicine* and *Disability, Health and Happiness in the Shakespearean Body*, and I would like to thank the reviewers and editors of those texts for their helpful comments and suggestions. I am also grateful to the Philadelphia Museum of Art, the Blanton Museum of Art, the Museo polo Firenze, the Wellcome Library and Manchester University John Rylands Library for permission to reproduce images from their collections.

Finally, I would like to thank my family and friends for their support through good and stressful times. Emma, Matthew and Sam provided, in no particular order, late-night proofreading, an ancient historian's perspective, tea, sympathy and motivational songs. Thank you.

Referencing Conventions

'ū' characters have been resolved to 'un' and y^e and y^t contractions expanded to 'the' and 'that'. 'V' and 'u' characters have been modernised.

Original spellings and punctuation are maintained, with the exception of 'hee', 'shee' and 'itt' (now 'he', 'she', 'it') and unnecessary spaces (for instance, before ';' and ':'), which have been removed. Long titles have been curtailed. Where a primary text does not contain page numbers, signature marks are given.

All references to Shakespeare are taken from William Shakespeare, *The Complete Works* (Second Edition), ed. Stanley Wells, Gary Taylor, John Jowett and William Montgomery (Oxford: Oxford University Press, 2005). Unless otherwise stated, all references to the Bible refer to the Authorized King James Version (Oxford: Oxford University Press, 1997). For plays, act and scene numbers are given in-text; line numbers are given for longer poems, but not for sonnets and short verse.

Introduction

On an unknown date in the mid-seventeenth century, Mrs Townsend, of Alverston in Gloucestershire, steeled herself for a dangerous but potentially life-saving operation. Mrs Townsend had breast cancer, and she was to have her breast 'taken off' by two surgeons, Mr Linch and Mr Clark. Watching the operation was Reverend John Ward, vicar of Stratford-upon-Avon. He recorded the events in his diary:

> They had their needles and waxt thread ready, but never ust them; and allso their cauterizing irons, but they used them not: she lost not above [six ounces] of blood in all. Dr. Needham coming too late, staid next day to see it opened. He said it was a melliceris, and not a perfect cancer; but it would have been one quickly. There came out a gush of a great quantitie of waterish substance, as much as would fill a flaggon; when they had done, they cutt off, one one bitt, another another, and putt a glass of wine in and some lint, and so let it alone till the next day; then they opend it again, and injected myrrhe, aloes, and such things as resisted putrefaction, and so bound it upp againe.
>
> Every time they dresst it, they cutt off something of the cancer that was left behind; the chyrurgions were for applying a caustick, but Dr. Needham said no, not till the last, since she could endure the knife...One of the chyrurgeons told her afterwards, that she had endured soe much, that he would have lost his life ere he would have sufferd the like; and the Dr. said he had read that women would endure more than men, but did not beleeve it till now.[1]

Little is known about Mrs Townsend, but her story raises some intriguing questions. How, for example, did the patient and her doctors understand 'cancer', and why was it deemed so serious that to be rid of it, Mrs Townsend was prepared to undergo major surgery in an age with neither anaesthesia nor antisepsis? What made the surgeons present believe that amputating the breast was the best course of action despite the 'suffering' it entailed, and why was that course so fascinating that both Ward and the eminent physician Walter Needham travelled to see it undertaken?[2]

This book examines these questions and many others in order to find out what cancer meant to early modern English men and women. It will contend that medical practitioners and their patients had a strong sense of cancer as a distinct disease which was marked out by unique pathological and zoomorphic 'behavioural' characteristics. In diverse sources, including poetry, drama, life writing, medical textbooks and medical practitioners' casebooks, cancer was constructed as fearsome and malign. Moreover, cancer was, unlike other serious diseases, conceptualised as both produced by the body and a hostile, independent parasite consuming that body from within. On one hand, the period's dominant medical model, that of the four humours, presented the disease as caused by physiological imbalances, particularly in the mysterious bodies of women. On the other, both medical and literary discourses imagined cancerous tumours as somehow sentient, eating up the body like a devouring worm or a ravenous wolf. In a bid to halt this deadly progress, medical practitioners found themselves engaged in increasingly dangerous and combative therapeutics, from toxic 'chemotherapies' to gruesome operations such as the one described above. In all, the concept and experience of cancer was moulded by, and in turn shaped, early modern people's patterns of thought in areas as diverse as the body, the medical profession, the state and gender attributes.

The study of early modern cancer is significant for our understanding of the period's medical theory and practice. In many respects, cancer exemplifies the flexibility of early modern medical thought, which managed to accommodate, seemingly without friction, the notion that cancer was a disease with humoral origins alongside the conviction that the malady was in some sense ontologically independent. Discussions of why cancer spread rapidly through the body, and was difficult, if not impossible, to cure, prompted various medical explanations at the same time that physicians and surgeons joined with non-medical authors in describing the disease as acting in a way that was 'malignant' in the fullest sense, purposely 'fierce', 'rebellious' and intractable.[3] Theories seeking to explain why cancer appeared most often in the female breast similarly joined culturally mediated anatomical and humoral theory with recognition of the peculiarities of women's social, domestic and emotional life-cycles. Moreover, as a morbid disease, cancer generated eclectic and sometimes extreme medical responses, the mixed results of which would prompt many questions over the proper extent of pharmaceutical or surgical intervention.

Knowing what cancer 'meant' also fills in a long-standing gap in readings of early modern imaginative and persuasive literature. When clergymen talked of the cancer of sin, or Shakespeare wrote of a 'canker ... in sweetest bud' ('Sonnet 35'), I argue that they accessed medical and somatic contexts which have hitherto gone unnoticed by literary scholars. Cancers, or 'cankers', connoted a specific set of characteristics: the ability to remain hidden or secret, the ability to spread rapidly through the personal or politic body and the likelihood of causing violent sufferings. Most significantly, 'cancer' signified a threat of which the origins were uncertain, both of the afflicted body and hostile to it. Constructions of cancer truly bridged the perceived gap between medical and cultural discourses and remain vital to a fuller understanding of both.

Finally, while this book is firmly rooted in the past, it may also contribute something to our understanding of twenty-first-century constructions of cancer. Medical perceptions of the aetiology and pathology of cancers have changed almost beyond recognition – as, mercifully, have treatment methods. Nonetheless, parts of my study will strike a familiar note. Notions of cancer as a purposely evil and cruel disease, or as a creature inhabiting the body, still seem to inform campaigns such as Cancer Research UK's 'Cancer, we're coming to get you'.[4] In the words of Ellen Leopold, one of the most prominent 'biographers' of cancer in the twentieth century, 'our habits of mind still betray the presence of age-old impressions and representations of the disease'.[5] As this book will explore, our collective fascination with and fear of cancer is nothing new.

I.1   Contexts: early modern medicine

In the period covered by this book, 1580–1720, understandings of cancer were situated within a medical landscape that is in many respects unrecognisable to the modern reader. Disease was predominantly understood, in theory at least, as a matter of individual bodily imbalance rather than exposure to distinct pathogens, and those whom one might consult for a diagnosis or cure varied widely, from the university-educated physician to members of one's own household.

Most of the primary material for this book is taken from medical textbooks created as instructional aids or thinly veiled advertorials by 'authorised' physicians, surgeons and apothecaries – that is, those who were members of the Royal College of Physicians, the Company of Barber-Surgeons or, after 1617, the Worshipful Society of Apothecaries. Also visible, however, are diagnoses and therapies from interested gentlemen and women, midwives, an array of apparently 'unauthorised' sellers of cure-all medicines and intriguing figures such as the 'un-born Dr', a 'monstrous' and seemingly unlicensed London surgeon.[6] Recent studies of the early modern medical marketplace suggest that such diversity was not unusual.[7] In London, though markedly less so outside it, a broad range of medical practitioners existed to suit most tastes and pockets, creating a more complex marketplace than simply 'authorised doctors' and 'quacks'. 'In reality', argues Andrew Wear, 'not only did lay people, empirics and others constitute important medical resources...but the occupational distinctions set up by the physicians were often ignored'.[8] University-educated physicians were less likely to practice outside major towns and cities, and therefore 'surgeon-physicians and apothecary-physicians...were common in the provinces'.[9] In addition, a thriving tradition of household physic blurred the boundaries between professional and amateur, with practitioners recreating medicines prescribed by the physician in domestic receipts of extraordinary complexity and potency.[10] Indeed, Ward's interest in Mrs Townsend's operation extended beyond human sympathy. The Reverend, who had a lifelong interest in physic and anatomy, frequently provided medical care to his flock, and even undertook minor surgery.[11]

Despite the abiding multiplicity of medical practice, it is clear that great efforts were made by licensed practitioners to stamp out certain areas of what they deemed quackery, and that these efforts only increased during the seventeenth and eighteenth centuries.[12] While physicians and surgeons were prepared to accept that freely provided household physic might be beneficial to those unable (geographically or financially) to access an authorised medical practitioner, those 'empirics' who charged for their services were often viewed with contempt.[13] These practitioners, it was claimed, undermined the work of authorised physicians, surgeons and apothecaries by offering gentler, more pleasant medicines. They also professed 'spurious foreign credentials', and sometimes advertised their remedies as rare cure-alls, with the aid of foreign jargon, exotic animals or costumes.[14] Empirics were presented as an omnipresent threat in discussions of cancer in medical textbooks, which, as Chapter 5 relates, told tales of terrible cancerous ulcers caused by the mismanagement of benign tumours. However, it was not only those outside the medical establishment who caused anxiety. In the seventeenth and eighteenth centuries, power struggles raged between (and within) the professional bodies of physicians, surgeons and apothecaries, each of which felt that they ought to be afforded greater professional status, and jealously guarded their tenuous monopoly on certain areas of practice.[15] In this environment, it seems that women wishing to practise medicine for money fared particularly badly. In my primary texts, there are relatively few women who made their living from medicine, and this reflects the assertion of numerous scholars that effectively, though not always legally, women were excluded from practising physic and surgery, and that their established role as midwives arguably diminished over the course of the seventeenth century.[16]

Whoever might administer it, the majority of early modern medical practice was underpinned by one theoretical model: the system known as 'humoralism' or 'Galenism'. In brief, this model was founded on the belief – outlined by Hippocrates, and expanded by the Greek physician Galen of Pergamon – that the body contained four humours which were associated with four combinations of temperature and moisture. Phlegm occupied the cold and wet corner of this spectrum, blood the warm and wet, choler (yellow bile) the hot and dry, and melancholy (black bile) the cold and dry. These humours circulated through the body in the nutritive blood (as distinct from 'pure blood', the sanguine humour) and lymphatic vessels. They also permeated tissues and organs, with some parts of the body having particular associations with certain humours. In the humoral system, the ideal human body was one which contained all four humours in their proper quantities. In practice, however, it was believed that this balance was virtually impossible to achieve, and through a combination of environmental factors and natural predisposition, most people tended toward one of the four 'complexions': phlegmatic, sanguine, choleric or melancholy. As Chapter 2 details, there was also a gendered aspect to this theory: the full range of such complexions was available to men, but women were, for various reasons, thought to be confined to the 'cold' end of the humoral spectrum. Complexions influenced nearly all aspects of physical and psychological health. They determined a person's ideal diet and susceptibility to certain diseases and shaped their emotional and mental predispositions, leading to the unique understanding of physiological and psychological phenomena discussed later. Unsurprisingly, therefore, explanations of the operation of the humours were often complex. The body's delicate balance was, Galenists believed, constantly influenced by both 'naturals' – humours, complexion, morphology and other things intrinsic to the body – and 'non-naturals', including sleep, exercise, environment, diet, climate and emotional state. This complexity, along with Galenism's emphasis on the need for anatomical training, was frequently the basis upon which physicians expounded the need for medical practitioners to possess a university degree, and decried the activities of so-called empirics.

Galen's influential medical writings frequently noted the author's debt to earlier physicians and philosophers, most notably Hippocrates.[17] In turn, as I will argue throughout the book, early modern interpretations of humoral medicine often showed their authors to have a keen sense of the extent to which their profession relied on pedagogy. Older practitioners advertised their texts as providing advice to younger fellows, and all drew on both ancient texts, from the likes of Galen, Celsus, Erasistratus and Aristotle, and medieval works, from continental practitioners such as Guy de Chauliac, Henri de Mondeville and Theodoric Borgononi. Thus, though medicine was always a dynamic field, it seems that, as Nancy Siraisi asserts, 'no sharp break separates [medieval and early Renaissance] medicine...from that of the early modern world'.[18] While it relied heavily on ancient and medieval texts, however, this period's medical practice was by no means devoid of new ideas.[19] In particular, much has been written in the past two decades on a supposed shift during the seventeenth and eighteenth centuries away from Galenism, and toward iatrochemical theories and therapeutics such as those proposed by the Dutch physician Jean Baptiste van Helmont and the famous Swiss physician, alchemist and occultist Paracelsus.[20] Paracelsus, and those who followed his method, rejected the teachings of Aristotle and Galen in favour of new observations of, and experiments with, chemicals; in particular, the *tria prima* of salt, sulphur and mercury, which together were believed to account for all physical properties. Accordingly, they held that diseases had material substance and could enter the body as 'seeds' which disrupted the local life force, or 'archeus', of a particular organ. The archeus would thus be prevented from operating in its usual manner to effect the unification or separation of substances within the body (the breakdown of food, for example), and disease symptoms would result.[21] Helmont's theory was of a similar bent, arguing that bodily processes such as digestion and respiration were essentially chemical in nature.[22] He too identified 'archei' at work within the body, which could be incited to 'fury' by disease seeds, extremes of emotion or bodily accidents.[23] Paracelsus and Helmont both presented themselves as revolutionaries, and their medical models as antidotes to a heathenish Galenic system practised by avaricious and corrupt physicians.[24] In contrast to their seemingly modern idea of diseases as ontological entities, however, both theorists also strongly believed in the influence of celestial or mystical forces on the body and 'envisioned a world full of occult energies'.[25]

Despite the radical potential of iatrochemical models such as those proposed by Paracelsus and Galen, recent studies have emphasised continuity, not change, in early modern medical practice. Numerous scholars have argued that iatrochemical medicines, and ontological perceptions of disease, did not suddenly revolutionise the sixteenth-, seventeenth- and early-eighteenth-century medical marketplace, but were rather incorporated into a medical landscape which remained broadly Galenist.[26] As Lindemann argues, 'Galenism endured because it was pliant and because its adherents were clever in weaving seemingly contradictory ideas and discoveries into its fabric'.[27] Just as the medical marketplace accommodated a variety of practitioners, Galenism avoided obsolescence by expanding to incorporate aspects from other medical theories, privileging the useful over the theoretically correct. By doing so, it remained influential in academic medicine well into the eighteenth century and culturally relevant for much longer. I shall refer to this synthesised, accommodating variety of humoralism at points throughout this book using the terms 'neo-Galenism' or 'neo-humoralism'.[28] As this book will show, uneasy alliances between new and old, authorised and empiric, professional and domestic were all to prove crucial to understandings of cancer.

I.2   Historiography

In the past two decades, the development of internet repositories such as *Early English Books Online*, *Defining Gender* and *Eighteenth Century Collections Online*, along with curated projects such as *Constructing Elizabeth Isham*, has increased almost beyond recognition ease of access to both printed and manuscript materials from the early modern period.[29] Accordingly, scholarship on somatic experience in this period has expanded considerably, and in literary studies, substantial attention has been paid both to non-canonical textual genres and to the positioning of aspects of canonical works (in particular, those of Shakespeare) within medical contexts. Of particular influence upon this book have been two overlapping modes of study: that which highlights the unique relationship between physiological and psychical well-being implied by the humoral model of the body, and that which traces the history of a particular illness, in which cancer is arguably underrepresented.

Since the 1990s, scholars of medical history and literature have increasingly turned their attention to considering how fundamental the humoral model might have been to early modern people's self-perception, and particularly to understandings of the relationship between psychic and physiological phenomena – or more broadly, the significance of bodily 'metaphors'. Here, I discuss the methodology of this book in relation to debates on illness and social constructionism. However, it is clear that humoralism also created a historically specific iteration of the cultural 'construction' of bodily experience. Medical and literary historians' approach to the 'figural/literal cusp' has been far from hegemonic but is consistently underpinned by the observation that in early modern understandings of the body, physical and psychological states were intimately and materially linked.[30] As Gowland observes, '[T]he advent of an emotion in the soul created a surge of its qualitatively corresponding humour to the heart'.[31] Feelings of anger, for example, provoked an increase in choleric humour, which in turn heated and agitated the brain. Body and mind operated upon a dynamic circuit, such that, it is argued, early modern people might have thought less in terms of a 'self' residing within the body and more of somatic, mental and spiritual experiences as interconnected and indivisible.

Moreover, interconnectivity was built into humoral theory, even down to the morphology of the human body. Proponents of Galenism argued for the existence of three 'venters' corresponding to the digestive organs, heart and lungs, and brain, and associated with the natural, vital and animal spirits, respectively. All three varieties of spirit, or 'pneuma', were necessary for human life, and all were influenced by the organs in which they circulated or were generated. The practical ramifications of this relationship between physiology and psychology were diverse. For example, it was popularly believed that maternal longings might imprint themselves onto an unborn child.[32] Certain conditions, such as lovesickness, were believed to cause physical changes to the brain and body which then exacerbated emotional distress.[33] In addition, as Jan Frans van Dijkhuizen and Karl A.E. Enenkel argue, a holistic, humoral model of selfhood could arguably alter one's most basic perception of bodily phenomena:

> Even evocations of physical pain that we would now tend to see as metaphorical, for example in descriptions of emotional pain, would have struck many early moderns as literal [...] Early modern culture construes intense emotions as inherently physical; their physicality even serves as an index of their intensity.[34]

Holistic understandings of the early modern body thus clearly influenced the experience and treatment of illness at a basic level. As Chapter 4 of this book details, they also contributed to the tendency to compare natural with politic bodies, and *vice versa*, a phenomenon which has been described in various permutations by medical, cultural and literary historians.[35]

Among the products of the 'bodily turn' in early modern studies have been a number of works focussing on specific illnesses, which often foreground the twinned physical and social ramifications of a particular disease. Venereal pox and plague have proven particularly fruitful topics for such investigations, with numerous authors showing how those diseases interacted with contemporary concerns about personal morality, national security and self-sufficiency.[36] Perhaps because it appears much less frequently in the primary literature, no such interdisciplinary study has been conducted of cancer in the early modern period. Indeed, while the politics and semiotics of cancer have been much studied, these studies overwhelmingly focus on the twentieth and twenty-first centuries, often from an activist standpoint. Texts such as *The Breast Cancer Wars* and 'The body in breast cancer', for instance, have shown how militaristic metaphors popularised in the mid-twentieth century continue to influence clinical research and decision-making around breast cancer.[37] Furthermore, cancer is, for many such studies, a feminist issue, with diagnosis and treatment for breast cancer in particular reflecting the 'pinkwashing' dominance of heteronormative models of femininity.[38]

In the related genre of cancer 'biographies', the recent past is equally, and understandably, emphasised. Several of the most incisive studies of the cultural and social history of cancer have concentrated on the twentieth century, and while they acknowledge the 'atavistic' presence of premodern beliefs about cancer in those narratives, these older beliefs are cast as static, homogenous and characterised by shame and fear.[39] Texts such as Siddhartha Mukherjee's popular *The Emperor of All Maladies*, James S. Olson's *Bathsheba's Breast* or George Johnson's recent *The Cancer Chronicles* offer a broader historical sweep, but nonetheless devote the vast majority of their pages to detailing the development of therapies in the past 200 years, an era of relatively rapid development in the understanding of cancers.[40] In many readings, therefore, cancer has been framed as a post-industrial disease, suddenly emerging as a major cause of death during the nineteenth century. Nevertheless, scholarship on cancer which traces the disease into pre- or early modernity has generally accepted that the disease is an ancient one, with textual evidence of 'cancers' dating back well over a millennium. A brief 2004 study by A. Kaprozilos and N. Pavlidis, for example, details treatments for the disease from the third-century BC writings of Hippocrates.[41] Others have antedated the disease even further, variously locating the first mention of cancer in the Edwin Smith Papyrus, an ancient Egyptian medical text thought to date from around BC 1500; the Indian epic *Ramayana*, BC *c*.2000; and the cuneiform tablets in the library of King Ashurbanipal of Assyria (BC 699–626), also thought to be copies of originals from around BC 2000.[42] Such scouting for 'original' cancers is a methodologically fraught exercise, since it often involves venturing into retrodiagnoses based on the application of 'correct' modern knowledge to disorders experienced in entirely different cultural and social contexts. Notwithstanding these pitfalls, such investigations have made clear that the ancient Greek understanding of cancer or 'karkinos', on which medieval and early modern scholars based their discussions, was probably not an entirely new disease categorisation.

While the antiquity of cancer is broadly agreed upon, its intervening history remains obscure. Whether cancer was recognised in Roman or Anglo-Saxon Britain is unknown, and the disease only re-emerges from the scholarly void in the medieval period. Several historians of medicine have briefly noted the inclusion of advice about cancer in medieval medical textbooks.[43] The most detailed study of cancer in the medieval period, however, and one to which I will return throughout this study, is Luke Demaitre's 'Medieval Notions of Cancer: Malignancy and Metaphor'.[44] Demaitre finds understandings of cancer in the medieval period to have been similar in many respects to those which I shall delineate for the sixteenth, seventeenth and early eighteenth centuries. Theories of the disease's causation were, he argues, mainly humoral. The malady was recognised by certain distinctive visual symptoms, and was accepted as usually fatal. Above all, Demaitre recognises that cancer was conceptualised in 'dramatic' terms as a 'subversive' illness, a theme which I will argue was developed in early modern discussions of cancer's pathology.[45]

Scholarship on the conceptualisation of cancer in the early modern period had, until recently, been more limited in scope. Both Wendy Churchill and Michael Stolberg have briefly described the most common symptoms of and treatments for breast cancer in this period.[46] From a literary perspective, Sujata Iyengar's *Shakespeare's Medical Language* has also lately focussed on 'canker' as a term which denoted cancerous disease as well as horticultural blight, and she briefly describes typical symptoms of the disease, as well as noting the use of 'canker' in the plays and sonnets.[47] Undoubtedly the most comprehensive work on early modern cancer to date, however, is Marjo Kaartinen's *Breast Cancer in the Eighteenth Century*.[48] Kaartinen's text discusses the supposed causes and methods of diagnosis for cancer, but focuses in particular on breast cancer therapies, both pharmaceutical and surgical, and on the physical experiences of women undergoing these treatments. She argues that breast cancer therapies underwent significant change during the latter half of the eighteenth century in particular, with mastectomies becoming more radical and invasive, and non-surgical remedies drawing on a range of exotic ingredients. Kaartinen's work is referenced at points throughout this book, particularly in my discussion of cancer treatments. Nonetheless, her text differs from my own in several respects. *Breast Cancer in the Eighteenth Century* focuses, for the most part, on a period later than that examined in this book, and Kaartinen's approach to cancer emphasises scientific innovation, particularly in the later eighteenth century, while paying relatively little attention to the disease's Galenic 'heritage'. By contrast, the chronological range of this book (1580–1720) is in my view characterised by relatively consistent views on cancer, underpinned by medical theory and praxis which remained predominantly humoral in character. Moreover, this book dwells less upon the physical experience of cancer than the ways in which its characterisation and representation shaped, and was shaped by, somatic realities.

I.3   Materials and methodology

My own interest in constructions of cancer during the early modern period was first aroused by the 1700–03 *Diaries* of Lady Sarah Cowper.[49] This remarkable, formidable woman had on several occasions documented her fear of getting cancer, the incidence of the disease among her friends and acquaintances and her own speculations on the causes thereof. Cowper's writings appeared carefully crafted, despite their ostensibly closeted nature, and presented an apt object for literary study. However, it was also clear that in order to read such writings, one needed to understand their historical context. Why, for example, did Cowper believe that a bruise to her breast might cause cancer, or that the uterine cancer of her acquaintance was caused by a 'foul' venereal disease?[50] In order to understand how early modern people thought about and experienced cancerous disease, this book reads medical texts and life writing through the lens of the literary scholar, and approaches literature as refracting and reshaping somatic experience. Furthermore, it contends that somatic and cultural experiences were not cleanly divided. In both literary and medical texts, how cancer felt, and what was said about it, were two sides of the same coin.

This approach is indebted to the work of numerous scholars of literature, history, and cultural studies. Still, the thorny issue of what exactly constitutes 'the body' is negotiated rather than resolved in the coming chapters. The thoroughgoing social construction of the body as posited by Judith Butler – that is, the insistence that there is no epistemic 'anchor' outside of discursive creation – seems, in the context of this book's subject, unfairly to deny the felt reality of pain and physical degeneration.[51] As Laura Gowing points out, 'knowing that the body is a product of culture does not tell us much about how it felt'.[52] I am conscious that behind the texts examined in the coming pages are a multitude of early modern people who almost certainly did not consider their pain, debility or bereavement as products of discourse. However, if, as Robert Aronowitz suggests, one starts from the premise that disease experiences are *contingent upon* discursive construction, then we can approach a more useful theoretical model.[53] This model still resists the idea that culture and metaphor get in the way of essential 'truths' about disease.[54] Rather, it suggests that social experience is embedded in, while not entirely constitutive of, experiences of the body.[55]

A broadly social constructionist model of bodily experience may be particularly useful when we are faced with an unfamiliar mode of thinking about that body. Shigehisa Kuriyama elegantly expresses this challenge in relation to the divergence of Greek and Chinese medicine:

> My argument is not about precedence, but about interdependence. Theoretical preconceptions at once shaped and were shaped by the contours of haptic sensation. This is the primary lesson that I want to stress: when we study conceptions of the body, we are examining constructions not just in the mind, but also in the senses. Greek and Chinese doctors grasped the body differently – literally as well as figuratively. The puzzling otherness of medical traditions involves not least alternate styles of perceiving.[56]

This book attempts to meet the challenge of an 'alternate style of perceiving' in several ways. First, it eschews the notion that medical history describes progress toward an 'enlightened' modern age in favour of a more complex narrative, which embraces the contingency of medical beliefs upon non-scientific factors. In this book, I will argue at various points that discussions of cancer from 1580 to 1720 show little sustained change. Though they became more numerous during the course of the seventeenth century, descriptions of cancer and its treatments returned time and again to the same images of hope and fear. In almost every chapter, there are examples of texts from the late seventeenth or early eighteenth centuries which closely echo those of the 1580s, 1590s and 1600s.

Secondly, the importance of cultural to somatic experience described here provides the basis for this book's unequal emphasis on certain aspects of the construction and experience of cancer. Cancer surgery, for instance (the subject of Chapter 6), appears to have been a relatively unusual way of treating the disease. However, it loomed large in both medical and non-medical discussions of cancer and possessed an importance to the conceptualisation of cancerous disease which outstripped its actual therapeutic use. In this book, I use the tools of literary analysis in order to highlight such points of anxiety or dissonance in textual representations of cancer. Thirdly, thinking about the cultural mediation of disease encounters has led me to reject, as far as possible, attempts to retrodiagnose cancer. Much literature on this subject has contended that certain examples of cancer found in the primary literature on this subject were misdiagnosed, perhaps from benign tumours or intractable cases of mastitis.[57] Elsewhere, symptoms, such as worms found in cancerous ulcers, which were presented in the primary material as intrinsic to cancerous disease, may appear to modern readers as 'really' a secondary complication. For the purpose of examining constructions and experiences of cancer, however, such diagnoses are anachronistic and often unhelpful. Bodily phenomena which were accepted in the early modern period as indicating cancers are treated as such in this book.

In addition to such theoretical influences, the methodological approach of this book has been more pragmatically determined by the unique set of materials upon which it is based, which are wide-ranging in terms of periodicity, geography and genre. First, the book covers a relatively wide period – 140 years – which has been chosen for a number of reasons. The seventeenth century, as detailed earlier, provided a melting pot in which humoralism met and melded with iatrochemical theories. The number of medical practitioners grew over this period to cater to an expanding population, and the activities of those practitioners became better-recorded as various factors combined to ensure that more texts were printed and kept for posterity.[58] The era also saw seismic shifts in the political and religious landscape, which were productive of much polemic, drama and poetry concerning the national 'body'. However, none of these changes can be viewed in isolation. To put the construction of cancer into its proper context, this book looks back to the late sixteenth century; the point at which the number of medical texts and medical practitioners seems to have begun a significant expansion, and at which enough texts start to survive to build up some picture of a relatively uncommon (or uncommonly diagnosed) disease as interpreted in different contexts. Looking forward, to the beginning of the eighteenth century, one can learn more about the appeal of early modern models of cancer by studying how those models underwent or resisted alteration as the empiricist medical theories of the Enlightenment began, tentatively, to take hold.

The book's geographical reach is less clearly defined. It explores the experiences of medical practitioners, patients and lay people in England, and is most concerned with texts published in England in the vernacular. These experiences and texts, however, were shaped by influences from mainland Europe and beyond. As detailed earlier, many of the most influential writings on cancer were translations from French, German or the European *lingua franca*, Latin. These relate cases and procedures which took place outside England, but they are included because, in translation, they became inseparable from English consciousness and practice. Most physicians of the early modern period could read Latin – indeed, it was at various points a requirement for admittance to the Royal College of Physicians and the College of Barber-Surgeons – but I have found that sustained discussions of cancer more frequently occurred in the vernacular, perhaps because the authors were keen to be associated with a modern, democratic style of medicine, or because such texts were of substantial interest to midwives and apothecaries, for whom Latin was not a prerequisite. In either case, accounts of cancer and its treatment from the continent show many more similarities to than differences from their English equivalents.[59] This is unsurprising given that many English physicians and surgeons had received either practical or academic training in France, in Germany or in the Netherlands.[60] In addition, medical practitioners from many parts of the continent could be found practising, and publishing, in England.[61] Within the British Isles, this book is often London-centric, and makes no reference to Ireland, Wales and Scotland. This reflects the contemporary bias in both texts and practice: London far outstripped the rest of the country in terms of population and concentration of medical practitioners during the early modern period, and although cases were recorded from other parts of England, and from France and the Netherlands, Ireland, Wales and Scotland were almost never mentioned in texts discussing cancer.

In seeking to trace cancer's cultural development, I have looked to diverse kinds of texts; principally, literary (poetic, dramatic, religious and polemical), medical and life writings. This reflects the degree to which it seems that seventeenth-century readers omnivorously consumed texts from the arts, sciences and philosophy. For much of the seventeenth century, 'science *was* knowledge', and *scientia* of the physical and metaphysical were not mutually exclusive.[62] Moreover, in places, I have deliberately juxtaposed the concrete – accounts of treatment, for example – with the abstract, in order to demonstrate the degree to which the same imaginative constructions of cancer informed both creative and practical reactions to the disease. Among the literary texts under my examination, political and religious polemic (in the form of poems, sermons and broadsheets) is particularly prominent. At the other end of the public-private spectrum, personal letters and diaries are treated in this book as both intimate forms of expression and crafted, persuasive works which were often intended for an audience, either in life, or after the author's death. With the juxtaposition of such 'literary' works with medical texts, however, come certain risks: most obviously, that of flattening contextual considerations, ascribing texts' differences or similarities to broad cultural trends rather than more localised economic, social or stylistic considerations. Brief details of these texts' pertinent economic and social contexts are, therefore, supplied here.

I.4   Modes of early modern medical writing

Most of the material in this book comes from the huge variety of medical textbooks published in the sixteenth, seventeenth and eighteenth centuries. These texts were diverse in authorship and intended audience, and I only detail here a few of the most prominent genres among my sources. As Furdell describes, it is difficult to discern precisely who was reading medical texts and why during this period.[63] Although some records of the contents of private libraries survive, many works were kept in coffeehouses to be read by the patrons, or were privately passed from one reader to the next.[64] Equally, while we can assume that texts which went through many editions, such as Nicholas Culpeper's *A Directory for Midwives*, were popular, we have little information on the numbers produced in each print run. In general, however, it appears that medical texts were a marketable product, especially as the seventeenth century progressed.[65]

A significant proportion of the medical textbooks examined in this book were authored by English, often London-based, medical practitioners, who were commonly, though by no means universally, licensed to practice by the Royal College of Physicians, the Company of Barber-Surgeons, or (after 1617) the Worshipful Society of Apothecaries. They frequently marketed the books as aids to the young scholar of medicine, while aware that the same texts would be of interest to gentlefolk with an academic interest in the subject. As well as general guides to the practice of physic or surgery, works abounded on individual procedures, life stages or illnesses. Works of 'advice' to midwives, mothers and wet-nurses were common, as were books of surgery, or texts dealing with the illnesses of certain (usually reproductive) parts. Many authors sought to make their name by focussing on an individual complaint; most frequently, plague or venereal pox, though tomes on various diseases from King's-evil to gout, leprosy and cancer could be found among London booksellers' wares.[66] Not only were such texts instructional, they conspicuously demonstrated the author's expertise in a particular area, often serving as thinly veiled advertisements.[67] Other medical practitioners presented texts which were similarly conceived as a mixture of instruction and self-promotion, but were explicitly targeted at lay people seeking to manage their own ailments, with titles such as *The Widowes Treasure*, which promised recipes suited to economy and common sense.[68] These were often aimed at women, who were understood to provide or oversee basic medical care and remedies to members of their household and, on occasion, the associated livestock. In many instances, they also dealt specifically with 'women's illnesses', with authors claiming that their books might help women to recognize their own ailments without medical consultations which might offend their 'natural' modesty. Once again, some of these texts advertised the author-practitioner or their remedies, with the cure for every ailment being a bottle of the writer's top-secret draught.

In addition to such general and disease-specific works, texts on pregnancy and childbirth were, unsurprisingly, among the most abundant in the medical marketplace, and feature prominently in this book. As Doreen Evenden observes, these texts provided a particular locus for debates about the proper role of women in medical publishing and midwifery more generally.[69] For instance, the 1698 edition of *The Compleat Midwife's Practice* possesses, as my Bibliography explains, a particularly convoluted authorial history, being first credited to four female midwives and later to four prominent male medical practitioners.[70] However, texts by women were not unheard of. The renowned midwife Jane Sharp, for example, was responsible for one of the seventeenth century's most popular books on pregnancy and childbirth, *The Midwives Book*.[71] Other women, such Alethea Talbot and Hannah Wolley, included medical receipts as a significant portion of printed texts on household management, building on the tradition of manuscript 'receipt books' as outlined below.[72] Still more women included medical advice in almanacs, like Mary Holden's *The Woman's Almanack*.[73]

The thriving British market for medical textbooks was also characterised by intertextuality and translation. The seminal texts of ancient authors such as Galen were virtually required reading for anyone claiming expertise in medicine, and were available in the vernacular, or in 'simplified' versions, in numerous editions from the mid-sixteenth century. Translations of more modern works came primarily from Europe, in particular, France, Germany, Switzerland and the Netherlands, and were usually rendered into English either by medical practitioners, or by unknown figures, seemingly in the employ of printers, who were often registered only by their initials. Different parts of Europe were at various times believed to have expertise in certain areas of medicine – Paris, for example, was known for surgery – and English readers eagerly consumed this expertise. By the eighteenth century, many continental textbooks were appearing in English translations only a year or two after their initial publication. Whatever their provenance, translated texts were probably coloured by the translator's own opinions, frequently featuring additions, amendments or marginal notes. Furthermore, all kinds of medical works borrowed freely from one another, often without crediting the author whose ideas they appropriated. In such circumstances, trying to discern what is 'original' work and what has been added is often an impossible task.

At the opposite end of the spectrum from published medical textbooks were receipt books, which offer a window onto the homemade remedies which often provided early modern people with their first (and sometimes only) means of defence against illness. These manuscripts often contained cookery and household receipts as well as medical remedies.[74] The receipts could be gathered from various places, including medical practitioners, friends and relatives, and receipt books bearing entries and amendments from numerous hands were frequently passed down the maternal line of families over many decades. As Chapters 1 and 5 will detail, these texts usually omitted any discussion of the theory of medicine or disease, simply recording those remedies which were 'probatum', or proven. This, along with their free use of medical terminology, makes them both valuable and frustratingly opaque sources for the modern scholar. Lastly, this project draws upon a small number of medical casebooks: texts which recorded, often in manuscript, a single medical practitioner's dealings with his patients.[75] Such texts offer a 'warts-and-all' insight into what treatments were actually prescribed for a complaint, and their effects. Casebooks demonstrate the process of trial and error by which diagnosis often took place, and the extent to which patients were treated as suffering from a compound of problems rather than a single complaint. Flattering examples from these collections were sometimes culled for inclusion in an author's printed works, while elsewhere, casebooks were published as stand-alone texts.[76] In either case, it seems likely that the practitioner substantially edited his or her notes prior to publication or production of a 'fair copy'. The detail (and legibility) of early modern casebooks is highly variable – some supply detailed case histories, whilst others contain brief notes of administered therapies, in abbreviations intelligible only to the writer. As part of the tissue of sources employed in this book, however, they offer a unique perspective on the difficulties of encountering cancerous disease.

*   *   *

This book is broadly divided into two themes. The first four chapters deal explicitly with beliefs about cancer, its symptoms, aetiology and 'character'. The last two chapters examine therapies for cancer, and how these shaped and were shaped by such beliefs. In Chapter 1, I establish some parameters for the book by asking, 'what was cancer?' Looking at the etymology and terminology of cancer, the diagnostic criteria for the disease and some of its supposed causes, I argue that cancer in the early modern period was a disease for which the pathological understanding relied on a holistic view of the disease's aetiology, prognosis, and perceived 'behaviour'. Such complaints, I will contend, were basically continuous with the malignant tumours we understand as cancers today, although the language in which such maladies were described differed from today's usage in several respects.

This theme is further developed in Chapters 2 and 3, where I look in more detail at how cancer was believed to operate within the body. In Chapter 2, I make the case that cancer was understood as a 'gendered' disease, primarily affecting the breasts of women, and ask why this should have been the case. Women's vulnerability to cancerous disease originated, I contend, in an understanding of sexual difference which was both physiological and social in character. That understanding was highly socially mediated, and women's supposed pathology was inseparable from their most distinctive social functions as wives and mothers. Accordingly, I contend, some medical practitioners and lay onlookers ascribed cases of cancer in women to factors including maternal nursing, emotional turmoil and domestic violence.

In Chapter 3, I analyse the ways in which cancer was associated with wolves and worms. As I demonstrate, cancers were often viewed as having ontological agency, devouring the body in the manner of a ravenous wolf or, in a more literal sense, a parasitic worm. This conviction sprang in part from prevailing cultural, religious and scientific discourses about worms and wolves which consistently positioned those creatures in relation to bodily and spiritual decay. In turn, I contend, belief in the 'creature-hood' of cancers, either in a literal or an analogical sense, materially influenced the somatic experience of, and medical approaches to, the disease.

Chapter 4 addresses what I shall contend was the defining character-istic of cancer in the early modern imagination – malignancy. In rela-tion to cancerous disease, I argue, this phenomenon was understood in its fullest sense, as denoting both a pathological characteristic and a broader cruelty or intractability. Looking first to medical explanations of the spread of cancer through the body, I examine some esoteric but illu-minating discussions which positioned cancer as poisonous or conta-gious. In the latter part of the chapter I show how medical and 'literary' or polemic texts operated reciprocally to construct cancer as a disease with social and cultural as well as medical meanings, which was under-stood by all parties as quintessentially 'evil'.

Finally, the last two chapters of the book look in more depth at the therapies with which early modern people attempted to stay or reverse the effects of cancerous disease. Chapter 5 deals with 'non-surgical' therapies, which are loosely defined as those which did not involve deliberately penetrating the skin. From recommendations for diet and regimen, through diverse animal and vegetable medicines, to applica-tions of mercury and arsenic, I argue that increasingly aggressive medical interventions for cancer gradually diminished the involvement of the patient in their cure, and instead foregrounded an adversarial relation-ship between the medical practitioner and a cancerous disease which seemed ontologically distinct from the person in whom it occurred.

This theme is continued in Chapter 6, which discusses surgery for cancer, and particularly mastectomy. I examine why patients might consent to this dangerous course, and what cancer surgery entailed. This therapy presented the ultimate opportunity for the patient to be rid of a cancer that appeared 'hostile' to their body, and for surgeons to prove the efficacy of their craft in 'defeating' a notoriously intractable malady. However, as I shall argue, surgery for cancer was also highly dangerous, painful and controversial. In the debates around cancer surgery, and the anxieties revealed by cancer surgeons' own accounts, one can detect both the deep-seated fear of cancer which drove such drastic interven-tions and medical practitioners' uncertainties over the proper limits of their craft.

1

What Was Cancer? Definition, Diagnosis and Cause

> CANCER, (in Surgery) a dangerous Sore, or Ulcer; as in a Womans Breast, & c.
> DEGENERATE CANCER, is one which succeeds an Obstinate or ill-dressed Imposthume.
> PRIMITIVE CANCER, (among Surgeons) is one which comes of it self.
> [...]
> CARCINODES...a Tumour like a Cancer.
> CARCINOMA...the Cancer before it comes to an ulcer.[1]

Published in 1721, Nathan Bailey's *Universal Etymological English Dictionary* demonstrates the complexity of early modern perceptions of, and terms for, cancerous disease. In Bailey's definitions, cancer slips between identification by its prognosis, origins and stage. Not everything that looks like a cancer is a cancer – 'Carcinodes' merely imitates that disease – but it is unclear on what basis one can differentiate between 'real' and false cancers, or spot a cancer in the first place. Moreover, Bailey's dictionary only scratched the surface of the variance seen in texts discussing cancer, which included differences in terminology and definition almost as numerous as those who wrote them down. The project of this chapter, therefore, is to determine *how* we should understand early modern cancer(s). Can we treat 'cancer' as a single disease, with a single name? What made this disease different from others with similar symptoms? By what other terms might it have been recognised, and how was it identified in early modern medical practice?

In the Introduction to this book, I noted that studies of the history of cancer have often taken a retrodiagnostic approach, applying modern medical knowledge to pre- or early-modern experiences of disease. This tendency has been most prominent in the common assumption that Medieval or Renaissance physicians and onlookers possessed a view of cancerous disease which was simply a less sophisticated version of that found in modern medicine, and that they made 'right' or 'wrong' decisions about diagnosis and treatment from that viewpoint.[2] Even in the latest and most comprehensive study of cancer in the early modern period, Marjo Kaartinen's *Breast Cancer in the Eighteenth Century*, the focus is firmly on the experience of cancer patients once they had been diagnosed, and as such, the author devotes only 4 of her 124 pages to examining the definition and diagnosis of cancers.[3] Departing from these treatment-focussed histories of cancer, I will argue that in the long seventeenth century, discussions of the etymological roots, cause, and symptoms of cancer were central to the discursive creation of the disease. Furthermore, these discussions took place in literary as well as medical texts.

To date, analyses of the meaning of terms such as 'canker' and 'cancer' in drama, poetry and polemic have been surprisingly few. One of the most in-depth discussions of the significance of 'canker', Jonathan Gil Harris's article on Gerard Malynes's 1601 *A Treatise of the Canker of England's Common Wealth*, focuses largely on the disease's connection to the canker-worm, and as such is detailed in Chapter 3.[4] Lynette Hunter, meanwhile, speculates on the meanings of 'canker' in *Romeo and Juliet*, and notes how, in that play, the Friar and the Prince 'deal with different kinds of canker: the canker that is the closed-over but ulcerous wound and the canker-worm that consumes the plant from inside its stem'.[5] While Hunter argues that both kinds of canker 'have the ambivalent potential to be at the same time internal contamination and external infection or contagion', she views medical 'cankers' as referring to ulcerous wounds in general, and thus overlooks the rhetorical potential of malignant *cancer*, of which ulceration was merely one symptom.[6] Sujata Iyengar's *Shakespeare's Medical Language* comes closer than Hunter's analysis to describing the full potential of 'canker' as a term which might describe several kinds of horticultural or bodily disease, emphasising the 'figurative implications' of a disease that 'kills or corrupts from within, sometimes unseen from the outside'.[7] Like Hunter, however, Iyengar views the 'canker' of an ulcerated wound and that of a malignant tumour as 'not readily distinguish[ed]' by early modern medical practitioners. In this chapter, I argue that despite lexical confusion between the two categories, the majority of printed medical texts did in fact show a clear understanding of the difference between 'cankerous' ulcers caused by wounds or complaints such as venereal pox, and the more serious disease of cancer.

As will become clear throughout this book, all aspects of the conceptualisation and experience of cancer, from diagnosis to treatment, were closely intertwined. Moreover, theories about the nature and causes of cancer were often uncertain and conspicuously incomplete. Nonetheless, this chapter examines three areas which we might think of as providing the basic framework for an understanding of cancer: discussions of what the disease should be called and why, opinions about where a cancer could occur in the body and what symptoms it might produce, and debates over the efficient causes of the malady. First, I examine the etymology of the term 'cancer' and how the disease of cancer was signified in language. The proliferation of early modern terms for cancer presents, as I discuss, both a challenge for the modern reader and a question over how far this disease can be imagined as a coherent concept. Equally, however, the rich etymological and linguistic 'life' of cancer contributed to the construction of that disease as a singular and unique malady. In the second part of the chapter, I look at the bodily locations of cancer – where it might occur on or in the patient – before outlining some of the most common markers by which this disease was distinguished from more benign lumps and bumps. Finally, I explore the ways in which cancer was imagined as a disease with complex humoral origins, based primarily in the much-maligned humour of melancholy, but often also associated with yellow bile (choler), and the burning or 'adustion' of natural humours into harmful and destructive substances.

1.1   Cancer or canker? The etymology and terminology of cancerous disease

What was cancerous disease called in the early modern period? As Bailey's multiple dictionary entries indicate, this question is more complex than it may first appear. Early modern medical practitioners used several different terms to refer to cancer. Some of these terms referred exclusively to the kind of malignant tumours and ulcers we might easily recognize as cancerous today. Others were less precise, sometimes denoting cancerous disease, and at other times referring to any variety of festering sore. Identifying the points of convergence and divergence between these terms is an essential first step in reconstructing beliefs about cancerous disease.

While early modern medical terminology was often bafflingly complex, terms for cancerous disease shared one clear referent. The most common names for the malady – 'cancer', 'canker', 'kanker' and 'chancre' – derive from the same etymological root: the Greek 'karkinos' (*Καρκινός*), or 'crab'. As I demonstrate here, many early modern writers discussing cancer were keenly aware of the term's etymology, and this creatural analogy was influential upon how early modern people diagnosed, and later treated, cancerous disease. Furthermore, it implied that cancerous tumours should be viewed as ontologically independent of the body in which they occurred. Intriguingly, though cancer terminology was unmistakably Greek in origin, it appears that Old English terms for cancerous disease similarly cast the malady as a discrete entity rather than systemic disorder. Pauline Thompson, for example, points out that in Old English, the term used for cancer matched that for the bite of a snake or spider, and the sting of a scorpion.[8] Writing on medieval understandings of cancer, Luke Demaitre also notes that

> the eating action became explicit in several vernaculars, including Old English. A Latin characterization of a cancerous ulcer as having 'taken away' (*assumpserat*) a patient's lips and nose was translated as '*cancor aet*.' *Bald's Leechbook* defined the disease with a simple synonymy, '*cancer pæt is bite*.'[9]

As Demaitre's observation makes clear, speakers of one or both languages seemingly recognised the correlation between a biting disease in Old English and a 'grabbing' disease in Latin. This stress on etymology as closely linked to pathology is visible elsewhere in medieval and early modern medicine.[10] For cancer, however, links between the terminology and the experience of cancerous disease seem to have been particularly strong, materially influencing diagnostic and therapeutic approaches to the malady.

With the meaning of the word 'cancer' so powerfully encoded in the disease's etymology, one might expect that identifying the disease in early modern writings should be a straightforward task. Unfortunately, primary evidence suggests that even for contemporary medical practitioners, this could become a complicated business. In 1684, for example, a translated work by the Swiss physician Théophile Bonet complained about practitioners using the term 'canker' too freely:

> The original of the Cheat and Errour is from hence; because *Theodorick* and *Lanfranc*, whom *Guido* [Guy de Chauliac] follows, distinguished a Canker, into a Canker an imposthume, and a Canker an Ulcer. The Canker an Imposthume is the disease so called by *Hippocrates, Galen,*

> *Avicenna* and others, rational Physicians and Surgeons: *But the Canker an Ulcer* (so *Guido* calls it) *is, when by reason of Ulcers or Wounds, irritated by sharp Medicines, bad melancholick humours become adust and troubled* ... But such Ulcers, though malignant, and often times stubborn, are not yet Cankers, nor ought to be confounded with a Canker, whose Contumacy far surpasses the Malice of all Ulcers.[11]

Bonet's complaint appeared to be about misdiagnosis. At its root, however, was the shifting terminology of cancer, which threatened to destabilise the disease category altogether. Bonet, like many of his contemporaries, used 'canker' instead of 'cancer'. His *Guide to the Practical Physician*, in which this quotation appeared, made abundantly clear that the disease described was identical with that pinpointed as cancer in other texts. Indeed, Bonet titled this section 'A Cancer, *or a Canker*'. Clearly, Bonet's 'canker' was merely a variant spelling of cancer which retained the ejective pronunciation from the Latin term, and it was to be used exclusively as such. The same can be said of many contemporary texts which refer to 'cancre', 'kanker' or 'cancor'. Confusion arose, however, because whereas 'cancer' almost always referred to the malignant disease as described throughout this book, 'canker' could, as Bonet complained, signify multiple conditions, of which malignant cancerous disease was only one. These included bodily ulcers and lesions of various kinds, mouth ulcers and venereal sores. As R.W. McConchie observes, this crucial distinction has not always been recognized in literary and medical history:

> The existence of an anglicized form alongside the neo-classical form hardly necessitated the desuetude and loss of the other, and the word in foreign form may still have a place in the lexicon. As is often the case pairs develop with differentiated uses, as with *cancer – canker*, and the omission of one of a pair from the *OED* helps to obscure this process.[12]

Where supplementary information about a disease is unavailable – as, for example, in many receipt books – negotiating between 'canker-cancer' and 'canker-other' can become a tricky business.

Outside the variations of 'cancer', 'canker' and 'cancre', a separate term was also employed by certain practitioners to describe cancers of the face in particular. Noli-me-tangere, or 'touch me not', was a phrase which played on the widely held belief that interfering with cancers made them worse, as discussed in Chapter 5. From at least the sixteenth into the early eighteenth century, a number of medical writers used the phrase alongside 'canker' or 'cancer': asserting, for example, that 'when [cancer] fixes on the Face, 'tis called a *Noli me tangere*, because that touching irritates it, and makes it a greater Ravage'.[13] Others, however, believed that noli-me-tangere was a disease similar or related to cancer, but not identical with it.[14] In the 1706 *Chirurgia Curiosa*, for instance, German medical practitioner Matthias Gottfried Purmann described noli-me-tangere as a disease which shared many of the characteristics of cancer, including the tendency to ulcerate, but was separate from and 'in some Particulars worse than a *Cancer*'.[15] Like 'canker' and 'cancer', this appellation for cancerous disease was intrinsically linked to its symptoms and prognosis. Unlike those terms, however, this phrase presents few challenges to the modern reader. Throughout the early modern period, discussions of the complaint consistently and clearly indicate whether the author uses 'noli-me-tangere' to denote facial cancers, or to signify a separate, though similar, skin complaint.

The terminological instability of cancer certainly presents a challenge to scholars. Nevertheless, it is clear that cancerous disease 'existed' in the early modern period, in the sense of there being a distinctive malady known as 'cancer' which was broadly contiguous with the illness sharing that name today. Early modern medical practitioners generally did not, like some modern physicians, view cancer as a host of separate diseases with similar symptoms. They understood that cancer could occur in different places, and be designated 'womb cancer', 'breast cancer' and so on, but they believed that the same mechanisms were at work in every case. Furthermore, medical writers' stress on the etymology of cancer indicated key directions in the development of the disease concept. By focusing on the crab, they gravitated toward a model of the disease as independent, even sentient. As I discuss here, they used the visual traits of that creature to establish a memorable shorthand by which cancer's most distinctive symptoms were easily recognized. Finally, the activities of the canker-crab promised a sinister and determined adversary, a disease that could bite and grab. Each of these characteristics was to prove influential in the early modern diagnosis, experience and attempted cure of cancers.

1.2   Symptoms and diagnosis

When, he, the sore hath searched, clens'd, and dressed,

With Tents, and Plaisters proper thereunto,

(And, all things els, befitting him to do)

If, on the Wound, his Medicine worketh nought
Of that effect, which, thereby hath been sought;
But, keepes it at a stand, or, makes it worse:
He, presently, begins another course;
And, if that, also, failes him, growes assured,
It is a Cancer, hardly to be cured[16]

In the mid seventeenth-century, at the height of national civil unrest, the poet and pamphleteer George Wither proposed a poetic *Opobalsamum Anglicanum* to soothe England's woes. The rhetoric underpinning his project, the 'Cure of Some Scabs, Gangreeves and Cancers Indangering the Bodie of this Common-Wealth', is discussed at greater length in Chapter 4 of this book. In this chapter, however, I wish to consider Wither's assertion that cancer could only be 'assuredly' known by its resistance to all forms of cure. This section looks at how early modern medical practitioners attempted to define cancer by describing its most recognizable locations and symptoms – and how they understood the disease as eluding or defying those efforts, presenting a shifting target of which the parameters could never reliably be established.

The question of *where* in or on the body cancer could occur was central to the diagnostic process. It presents, therefore, an appropriate starting point for examining how medical practitioners and lay people looked at and for this disease. Elsewhere in this book, I make the case for cancer as paradigmatically a disease of the female breasts. For various medical and cultural reasons, I argue, the 'dugs', and to a lesser extent, the womb, of nature's supposedly weaker sex were understood as uniquely vulnerable to this disease. Thoughts of cancer would have come far more readily to a medical practitioner examining, or a patient discovering, a lump in her breast than anywhere else on the body. However, although these locations loomed large in the pathology of cancer, they did not define it absolutely. While attention was certainly concentrated on particular 'cancer-prone' areas, it seems that, given sufficiently compelling symptoms, some medical practitioners were prepared to diagnose cancer in almost any external part of the body. In particular, the 'upper partes about the face, the nosethrills, the eares, the lippes' were identified as being at special risk.[17] Like the breasts, the soft flesh of the face was deemed vulnerable because of its 'glandulous and spongy' nature, which provided the perfect environment for sluggish humours to coagulate and thicken.[18] These tissues may also have been common sites of diagnosis for more pragmatic reasons. Facial tumours could not remain hidden for long, and even the staunchest sufferer would struggle to ignore the likely disruption to speaking, eating and breathing wrought by a large tumour or ulcer. In severe cases, facial cancers could spread widely, ulcerate and eat away at the patient's bones.

Producing painfully obvious symptoms which, sooner or later, forced sufferers to seek medical advice, it is clear that the vast majority of all diagnosed cancers were on or near the surface of the body, in the breasts, face and skin. Indeed, many early modern authors presented cancer as affecting only these areas. At various points throughout the early modern period, however, individual medical practitioners occasionally discussed and diagnosed cancer in the throat, tonsils, cervix and even the lower part of the intestine. This passage, from the prominent surgeon Richard Wiseman, outlines some of the challenges such diagnoses might pose:

> Cancers may also be said to differ as they affect several Parts of the Body, as the Head, Face, Eyes, Nose, the Palate, Tonsils, Throat, Tongue, Jaws or Lips...
>
> Cancers affecting the *Uterus* and *Podex* [rectum] may also be distinguished as they are in the interiour or exteriour parts...Those that possess the body of the *Uterus*, or the upper part of the *Rectum intestinum*, are not discovered till they have made some progress; in which cases there is a bearing down, with a suppression of Urine. [...]
>
> If they be ulcerated, a filthy *Sanies* will discover it. If it be in the *Intestinum rectum*, the difficulty and pain in going to Stool will be exceeding great. If the *Uterus* be cancerated, there will be Fever, nauseousness, anxiety of mind. In some of those who died so diseased I have opened the Body, and found the *Uterus* preternaturally big and hard: in cutting into it I hav[e] seen it all rotten, Those in the more exteriour parts, whether it be of the Womb or *Podex*, are sooner discovered, and the Patients are in a greater possibility of being eased of their pains.[19]

Wiseman's description demonstrates that even when practitioners were aware of the possibility of internal cancers, diagnosis depended largely on the cancers either producing externally visible corollaries (tumours around the anus, or fetid 'sanies') or being palpable by the examining practitioner. When cancer invaded the innermost, 'interiour' parts of the body, the impossibility of safely conducting investigative surgery made diagnosis overwhelmingly difficult. As such, tumours of the vital organs were hardly discussed at all, and those discussions were usually brief, pointing out the near-impossibility of either identifying or treating the condition in such circumstances.

Knowing where cancers might occur, how was one to discern this disease from the many other skin complaints to which early modern people were susceptible? Given that most cancers were diagnosed on or near the surface of the body, it is unsurprising that visual symptoms were most prominent in medical textbooks' descriptions of cancer, setting the stage for an abiding concern with the (in)visibility of this disease. From the 1580s into the first decades of the eighteenth century, medical practitioners consistently talked about the colour of cancerous tumours, which varied from an unspecified livid hue to 'blackish, and sometimes inclined to black and blue'.[20] Moreover, it was expected that cancer's livid appearance would accompany a distinctive shape to the tumour, which was both 'rough and unequall' and 'round'; that is, circular, but with an uneven surface appearance.[21] For medical practitioners writing about and encountering this disease, a round, highly coloured swelling was therefore an immediate source of alarm. Nonetheless, these were characteristics that could and did appear in other, more benign, growths – including undifferentiated 'cankers'. The most definitive of cancer's visual symptoms was one which medical practitioners presented as occurring solely in this disease, and which was taken not only as proof of cancer's presence but as a sign of its 'evil' nature. Darkened blood vessels spreading outward from the suspect tumour seemed to illustrate the spread of malignant matter into the surrounding flesh, and this sign recurred in medical texts across the early modern period as the preeminent visual marker of a dangerous cancer. In the 1587 *A Worthy Treatise*, for instance, cancer was said to be characterised by 'Veines swollen rounde about with melancholicke bloude'.[22] Over a century later, the 1698 edition of *The Compleat Midwife's Practice* similarly noted that breast cancer might be 'known by the crooked windings, and retorted veins that are about it'.[23]

These visual features were firmly established as essential to the diagnosis of cancer, having been common to texts on the subject since the medieval period.[24] Each one was also consistently reiterated, creating a consensus on the visual signs of a 'true' cancer that was remarkably stable compared to the vigorous debate which surrounded the disease's treatment. Such consensus relied partly upon medical writers' tendency to liberally 'borrow' from one another's work. However, it was also underpinned by the compelling narrative which united diverse visual traits with reference to the figure of the crab. Each of the signs noted hitherto was consistently and explicitly aligned with parts of the crab's body. For instance, the roundness of cancer and its colour were both compared with the creature's round and vividly coloured carapace, while the blood vessels extending from the tumour were 'verie like unto the feete of crabbes, descending from the round compasse of their bodies'.[25]

Visual symptoms were central to the diagnosis of cancerous disease, and images of the cancer-crab helped codify those symptoms into a vivid and memorable format. In addition, numerous texts identified pain – specifically, its presence, type and extent – as a deciding factor in distinguishing cancerous from relatively benign scirrhous or phlegmatic tumours.[26] As the German physician Christof Wirsung vividly described, 'the Canker causeth...great paine and beating, whereof *Schirrhus* is free'.[27] Others described an 'exquisite pricking' or 'corrosive, cruel and terrible pain'.[28] Often coincident with pain as a diagnostic criterion was a 'certaine straunge, and extraordinarie heate' believed to attend cancerous tumours.[29] Undoubtedly, medical practitioners' interest in heat as a symptom originated in part from Galenic doctrines which positioned health as related to bodily temperature, and to discussions of cancer's cause which pinpointed the 'burning' of melancholy humours as particularly dangerous. In these observations, one can also detect an imaginative fascination with bodily heat. Images of the blood 'in the veines growing hot' depicted the natural and 'vital' warmth of the healthy body transformed into something beyond regulation, for which the inevitable end seemed to be the chill of death.[30] Furthermore, the pains associated with cancer could, once again, be aligned with the crab. In 1597, for example, physician Peter Lowe asserted that not only did cancers look like crabs, they 'gnaweth, eateth and goeth like this fish'.[31]

The use of the crab image as a means of reinscribing the visual and sensory symptoms of cancer thus remained immensely popular throughout the early modern period. The success of this device, however, depended on something more than its fit to cancer's visual characteristics. As an animate creature, the crab lent itself naturally to one of the most defining and enduring characteristics of cancer diagnostics – the reading of this disease's symptoms as sentient behaviours. In 1583, physician Philip Barrough asserted that '[s]ome have given [cancer] this name [crab] because it is verie hardly pulled awaie from those members, which it doth lay holde on, as the sea crabbe doth, who obstinately doth cleave to that place which it once hath apprehended', while in 1635,

Read added that 'whatsoever it claspeth with the clawes, it holdeth it firmly...[so] that it seemeth to be nailed to the part'.[32] The grip of the crab was understood not only as painful but as immensely strong and tenacious, matching precisely the intractability and resistance to cure which was one of cancer's most distinctive features. A renowned French practitioner Pierre Dionis made the connection explicit in 1701 when he explained that "'Tis no more possible to extirpate [cancer], than force a Crab to quit what he has grasped betwixt his griping Claws', while in the sixteenth century, Paré deemed the link between the 'tenacity' of cancer and the 'toothed claws' of the crab so instructive that he inserted a picture of the creature into his writing on the subject, to drive home the 'perspicuous' nature of the comparison.[33]

In the figure of the crab, early modern medical practitioners effectively united the diverse visible and invisible symptoms of cancer. Moreover, this practice appears not to have problematized, or been problematized by, understandings of cancer as humoral in origin. This phenomenon is seen amplified in Chapter 3 of this book, where I discuss the casting of cancer as a type of worm or wolf. Although medical practitioners had a good sense of cancer's symptomatology, however, there remained an element of doubt in any diagnosis. As Wither's verse suggested, in order to really be sure that a patient was suffering from cancer, one had to see whether the suspect tumour followed the most distinctive cancerous 'behaviour', that of expanding and spreading throughout the body. Malignancy was, as I shall discuss, fundamental to the very meaning of this disease, setting 'true' cancers apart from the myriad of less dangerous ulcers and neoplasms. Furthermore, it presented a counterpoint to all medical writers' diagnostic criteria. The way to 'know' a cancer was to see it growing; however, that hardly required medical expertise, and once a cancer had grown large, it was much more difficult to treat. Diagnosis therefore presented the first of this disease's many challenges to medical wisdom. Encounters with suspect tumours were not only matters of clinical determination, but of defining human relationships to cancer.

1.3 Causes of cancer

By describing cancer's symptoms, and emphasising its crablike 'nature', medical writers sought to distinguish this disease from other tumours and ulcers. Just as importantly, however, these authors attempted to work out *why* some people got cancer while others remained healthy.

Speculation about the causes of cancer was primarily found in instructional medical textbooks, for several reasons. First, it was deemed important for students of physic and surgery to understand how their therapies affected the underlying causes of a disease. Secondly, some medical texts implied that a practitioner's distinction between cancer and diseases with similar symptoms could, and should, be made on the basis of the patient's humoral make-up, something which could be discerned through a raft of signs apparently unconnected to the cancer. John Browne, for example, encouraged medical practitioners to distinguish between cancer and the less serious disease of scirrhus (sometimes thought to precede cancer) by considering that 'a *Scirrhus* is made by natural Melancholy, which is in the Blood, as the Lee is in the Wine; but a Cancer is not bred from natural, but adust Melancholy'.[34] Maynwaringe went still farther, categorising a whole range of tumours, from *Phlegmon* to *Inflatio*, by their humoral cause.[35] Unusually, his discussion of tumours also dwelt upon internal tumours and the difficulty of their detection; in which scenario any clues offered by the patient's humoral complexion were particularly valuable.[36]

Writers discussing cancer tended to draw broadly similar conclusions about the origins of the disease. Overwhelmingly, and in line with early modern medical orthodoxy, medical practitioners emphasised the provenance of cancer as humoral. More specifically, the disease was believed to arise from the much-maligned substance of black bile, or melancholy, which turned into *atra bilis* under certain circumstances. Causes of an excess of black bile were numerous, but the humour's effects were well documented. 'Cold and dry, thicke, blacke, sowre', it provoked diseases including epilepsy, ulcers, paralysis and, most notably, the disease of melancholy or melancholia (for clarity, I henceforth use 'melancholia' to describe the disease of melancholic 'depression' and 'melancholy' or 'black bile' to denote the humour).[37] Although presenting a potential hazard for any early modern body, melancholy, and the maladies associated with it, were associated in particular with the elderly, since with age came a natural 'diminution of spirits and substance' which saw the body becoming colder and drier.[38] Women, as Chapter 2 of this book details, were thought to be naturally colder than men, and old women were therefore particularly at risk of melancholy complaints.[39]

While excess melancholy could pose a health risk in itself, the vast majority of medical texts did not identify the simple presence of that humour as cancer-causing. Rather, they surmised that it only worked real mischief when either confined to a certain area, transformed into a more harmful substance, or both. Medical practitioners' means of describing these phenomena were diverse, and often confused, but consistently centred upon images of congestion and heating which subverted the principles of balance and circulation underlying the Galenic model of good health. Robert Bayfield, for example, asserted in 1662 that 'when this melancholious humor, resembling in proportion the dregs of wine, doth descend and flow into any member, and there abideth compact together, it causeth sometimes the disease called Varices, and sometimes it breedeth a Cancer, as when the same is somewhat cool'd'.[40] Bayfield's comparison of melancholy humour with a waste product, the thickened dregs of wine, was one seen repeated in several other discussions on cancer during the period. In 1583, Barrough similarly wrote that that melancholy 'resembleth the dregges of wine, & the filthines of oyle', while in 1703, Browne noted that the humour was 'in the Blood, as the Lee is in the Wine'.[41] There was an obvious internal logic to these claims – since movement and vigour created (and might result from) bodily warmth, melancholy, which occupied the 'cold and dry' corner of the humoral system, was bound to lack those qualities. Certain physicians also linked the sluggish and viscous movement of melancholy to the dysfunction of organs elsewhere in the body, notably the spleen. While the exact role of this organ in the regulation of the humours was often unclear, writers of medical textbooks repeatedly cited 'the infirmity or weakenesse of the spleene in attracting and purging the bloud' as a cause of tumours.[42] According to Read, this connection was attributable to Galen, who posited that the organ somehow drew 'superfluous naturall melancholy' from other parts of the body, preventing the mischiefs associated with that humour dwelling too long in one place.[43]

However, the persistence with which melancholy was imagined in cancer texts as thick, dark, sluggish and potentially dangerous was not only a product of morphological theory. As Demaitre notes of the medieval period, the conceptualisation of melancholy as related to cancer also 'underscores the suggestive power of humoral physiology'.[44] Black bile possessed a well-established cultural and medical 'biography' by the early modern period. Angus Gowland notes that early modern ideas about black bile, and particularly its role in the generation of madness, were broadly continuous with those of medieval and ancient Greek texts.[45] Notably, black bile was also subject to the same sort of terminological instability that dogged cancer.[46] As well as describing a particular substance, or a constitution in which that humour dominated, 'melancholy' also described a disease derivative of, and yet conceptually different from, black bile. Indeed, in his work on early modern selfhood, Charles Taylor sees the relationship between black bile and melancholia as exemplifying the necessity of a historically specific understanding of the relationship between humours and the diseases they caused:

Melancholia is black bile. That's what it means. Today we might think of the relationship expressed in this term as a psycho-physical causal one. An excess of the substance, black bile, in our system tends to bring on melancholy. We acknowledge a host of such relationships, so that this one is easily understandable to us, even though our notions of organic chemistry are very different from those of our ancestors.

But in fact there is an important difference between this account and the traditional theory of humours. On the earlier view, black bile doesn't just cause melancholy; melancholy somehow resides in it. The substance embodies this significance.[47]

Taylor's claim echoes the observation of Robert Burton, author of the popular *Anatomy of Melancholy*, that it was almost impossible to say 'whether [melancholia] be a cause or an effect, a Disease, or Symptome'.[48] It also implies that the relationship between black bile and melancholia, or black bile and cancer, is more fundamental than one might imagine, such that black bile may be said to be the progenitor of both these diseases in an organic sense, imbuing them with its own material qualities. Thus, contemporary discourses about melancholia may have influenced discussions of black bile and its other resultant diseases – including cancer.

The properties associated with melancholy and melancholia were almost universally negative. Gowland, for example, argues that a burgeoning tendency in the seventeenth century to ascribe seemingly supernatural powers (such as those of witches) to the effects of melancholia relied in part on 'the common assumption that devils were analogically attracted to interfere with complexionate melancholics because of the dark and semi-excremental nature of the black bile predominating in their bodies'.[49] Similarly, in his discussion of the supposed hallucinatory effects of melancholia, Clark points out that '*balneum diaboli* (the devil's bath)' was a common moniker for melancholy humour.[50] Bridget Gellert Lyons asserts that melancholy's association with Saturn imbued it with certain 'crafty, envious, secretive...maleficent' moral properties, which were particularly useful to contemporary poets and dramatists.[51] It is easy to see how this information might colour one's reading of cancer, a disease which was itself consistently figured as evil.

Even for those writers who did not view melancholy as malign or devilish, the humour's characterisation as excremental positioned it as dirty and undesirable, a view upheld by Burton's description of melancholy as drawn from the 'faeculent part of nourishment'.[52] In her work on humoralism and cosmology, Gail Kern Paster shows how melancholy accordingly became a watchword for filthiness in drama and polemic as well as medical texts. 'In *The Terrors of the Night*', she observes, 'Thomas Nashe likens "the thick steaming fenny vapours" of bodily melancholy to waste water'. Just as stagnant puddles 'engendered' foul creatures, so melancholy bred monsters in the imagination.[53] For the reader of early modern medical texts, the tendency of melancholy to cause cancers by becoming blocked up or stagnating in a certain area was thus to some degree inherent in that humour's dirty, troublesome nature. However, there were further dimensions to the link between melancholy and cancer. Across the early modern period, but particularly from the mid-seventeenth century, printed medical texts consistently pointed to the 'adustion' (heating or burning) of melancholy humours as a crucial step in rendering those humours harmful in general and cancer-causing in particular. Browne, for example, asserted in 1703 that 'a *Scirrhus* is made by natural Melancholy...but a Cancer is not bred from natural, but adust Melancholy', while in 1635, Read drew a similar conclusion when he stated that cancers commonly appeared in late summer and autumn 'because in these seasons, the melancholick exceedingly increaseth, and humors become adust'.[54] Even while disputing the model, Gendron and Wiseman, both prominent medical authors and practitioners, grudgingly admitted that adustion had become the predominant theory on the generation of cancers.[55] What adustion actually comprised, and how it occurred, was less clear. Medical practitioners variously ascribed the process to the dysfunction of the liver or spleen, the influence of other humours, the native heat of the body, and external factors such as diet. Most often, as is visible in this passage from Read's *Chirurgicall Lectures*, they blamed a cornucopia of factors:

> There are sundry efficient causes which ingender these humors in our bodies: First, a strong hot distemperature of the liver, which burneth the naturall melancholy and yellow choler, and so hatcheth this *Bilis atra*. Secondly, according to *Galen*... the spleene by reason of its weaknesse and distemperature, doth not draw unto it selfe the superfluous naturall melancholy, and so staying long without its owne proper place it is inflamed and burned. Thirdly, sometimes this humor is caused of the menstruall courses, and Hemorrhodes stopped. Fourthly, verie often an ill diet breedeth this humor (...) An hot aire and perturbations of the mind set forward also this humor.[56]

The external factors – diet, amenorrhea and 'mind set' – identified by Read are discussed elsewhere in this book. In common with many of his peers, however, Read identified the causes of adustion with more certitude than specificity. In general, medical practitioners positing a humoral explanation for cancer looked only so far inward – to the level of adust melancholy or *atra bilis* – before, like Read, they turned their gaze once more toward the environmental factors which aggravated that substance. They were therefore either unable, or saw no good reason, to supply details of exactly what happened inside the body to turn melancholy into these more harmful substances. The neo-Galenic model seems not to have fostered inquiry into the mechanics of each humour's operation, but rather focussed upon their qualitative characteristics. One particularly interesting theory, however, which we can see fleetingly referenced in Read's 'burning of naturall melancholly *and* yellow choler', was that adust or poisonous forms of melancholy might either have been comprised of several different humours, or of a different humour – choler, for example – which mutated into melancholy during the process of adustion.[57] While this kind of 'compound' melancholy is not evident in most texts on cancer, it is present in a number of discussions of the malady's cause, where a posited link between adust melancholy and choler (yellow bile) often provides a logical bridge between the efficient causes and the characteristics of the disease.[58] These discussions occurred over the sixteenth, seventeenth and eighteenth centuries, and may have been derived from ancient writings, though this remains unclear.[59] In his 1684 *Adenochoiradelogia*, for instance, Browne asserted that 'when [cancer] takes Adust Choler into its cognizance, and this gains better and nearer acquaintance therein, this in time masters the other, and makes the Patient feel the Vigour of its prevalency, by its corrosive, cruel and terrible pain which it brings along with it'.[60] Authors who discussed 'compound' melancholy were clear on the fact that yellow bile changed the character of resulting diseases for the worse. 'Hot, dry [and] bitter', choler was associated with anger and fierceness, and in his 1621 *The Anatomy of Melancholy*, Burton pinpointed choler as the root of 'brutish', 'rash, raving' varieties of madness.[61] Moreover, Jennifer Radden notes that, according to Galen, yellow bile was associated with acute diseases and black bile with those of long continuance.[62] In theories of 'choleric' melancholy, therefore, one sees particularly clearly the marriage between discussions of cancer's cause and its troublesome, 'fierce' character, alongside a ready explanation of how the disease could be both acute in effects and chronic in duration. Furthermore, the language in which such correlations were described once again makes obvious how readily early modern people embraced emotive discourses of the fierce, filthy and mutable nature of certain bodily substances.

While these theories of adustion may have been lacking in some respects, they retained a largely unchallenged hold over how cancer was imagined until well into the eighteenth century. Iatrochemical language seeped into discourses of cause at various points: in particular, the 'bad' melancholic humour or *atra bilis* was often described as acidic or acrid.[63] However, the texts employing these phrases usually used them in conjunction with humoral ideas, seemingly seeking to lend gravitas to their conclusions by employing the newest terminology. In the period under my examination, only a handful of medical writers offered real alternatives to neo-Galenic theories of cancer's cause. Van Helmont's radical theories of disease causation have been well documented by critics and remained unaltered for cancer, positing the mysterious 'Archeus' as the agent of disease.[64] His approach, however, seems to have had little impact on the majority of medical practitioners or lay writers concerned with this disease. Elsewhere, Wiseman and Gendron provided visibly different alternatives to the above humoral models, but which remained linked to neo-Galenism. Wiseman, for example, scorned traditional ideas about adustion in his *Several Chirurgical Treatises*, scoffing that 'I cannot imagine what heat these Authors suppose to be in the Body which is capable of making such an Adustion as is here spoken of'.[65] He went on, however, to propose a model which integrated both humoral and iatro-chemical concepts, stating that cancer-causing humours were 'sharp and corrosive' because of some 'error in the Concoction' involving – though in a rather confused manner – 'acid Salts'.[66] Wiseman's near contemporary, Gendron, went even further, proposing that cancers were 'nothing else ... but a change of the Nervous Glandulous Parts, and the Lymphatick Vessels into an uniform, hard, close indissoluble Substance, capable of Increasing and being Ulcerated'.[67] That change, he insisted, was not a humoral one, but was caused by malfunction in the filtrative tissues found in those parts of the body affected by cancer.[68] As these tissues broke down and compressed into a lump, the vessels around them came under increased pressure, causing them to break down in turn, and so on. Both authors claimed that their models were based on extensive experimentation.[69] However, while their claims of scientific rigour may have reflected a medical community increasingly invested in the experimental principles of its work, neither author's purported objectivity prevented him from using the same highly emotive terms as were seen in emphatically humoralist texts on the genesis of cancer. Of the cancerous tumour, Gendron stated that 'Nature, if I may so say, is out of order', and continued the use of organic and even anthropo-morphic images in talking of a cancerous ulcer 'which ... destroys its own

Substance, by a Progressive Putrefaction'.[70] Similarly, Wiseman slipped into well-worn descriptions of cancer as anthropomorphically 'rebellious' and 'malign'.[71]

Clearly, the vast majority of writers on cancer adhered broadly to theories which positioned adust melancholy as the immediate cause of the disease. Even some of those who ostensibly rejected this model incorporated aspects of the theory into their alternative theses. What made this idea such an appealing and influential one, and how did it affect the perception of cancer's pathology more generally? As noted earlier, such theories accessed the wealth of imagery attendant on melancholy as part of both medical and broader cultural discourses. Moreover, *adust* melancholy offered solutions to a number of troubling aspects of the humoral model of cancer's causation. That is, it helped to explain why cancer patients frequently lacked any melancholic symptoms prior to the onset of their cancer, by arguing that patients suffered less from an excess of the humour than an accident in its formulation. It also avoided blaming serious illness on a substance which was supposedly natural and native to the body, as well as clarifying – either through the 'heating' or 'choleric' models – why these swellings, caused by a cold and dry humour, were often so hot to the touch.

As importantly, adust melancholy carried a cultural freight which expanded in many respects on negative beliefs about 'normal' melancholy.[72] This mid-seventeenth-century poem on 'Religion', for example, picked up the well-worn idea of black bile as the humour of witches and devils and reapplied that notion to adust melancholy in particular. 'Evill Spirits', wrote the author,

> have been, in Adust,
>
> Black Choler, sayd, to find a Tempting Gust
>
> (From whence their own Familiar-Imps, like Leaches
>
> Are Nursd, and Suckled, at the Teats of witches)[73]

Such suspicious attitudes toward adust melancholy were repeated in the loaded language of medical texts. The French medical practitioner Paul Dubé, for example, identified adust humours as 'nothing else than a natural Humour degenerated from its natural Disposition, and turn'd into a foreign form', adding that such humours proved particularly 'Malignant' and troublesome.[74] According to this rhetoric, adust melancholy was decisively alien to the body, having been utterly transformed from the sometimes harmful but ultimately native substance of ordinary melancholy. That concern was reiterated in Browne's assertion that 'Cancer is not bred from natural, but adust Melancholy': adustion was a product of which the organic genesis was implied in that term 'bred', but which was, like cancer itself, an unnatural progeny.[75] Bonet, citing the prominent medieval writer Guy de Chauliac as his influence, likewise summarised adust melancholy in emotive terms. '[B]ad melancholick humours', he wrote, 'become adust and troubled, and are drawn ... to that place, where they putrefy, grow hot, and acquire an acrimony and poisonous quality, whence there is an increase of the evil disposition, and it becomes a Canker'.[76] One sees in this passage the natural conclusion of the discourses positioning adust melancholy as 'unnatural': the casting of that humour as a poison, created by the body but now, like the cancer itself, hostile to it. Furthermore, the adustion of the humours marked, for Bonet, their transition from merely 'bad' to the anthropomorphic terms of 'troubled' and 'evil', sentiments which, as Chapter 4 demonstrates, were common among medical practitioners struggling to express the malignancy of the disease.

Beliefs about the humoral origins of cancerous disease played a crucial part in how cancer was imagined by both medical practitioners and lay people. Unsurprisingly, it also shaped therapeutic responses to cancer. As I shall discuss, humoral medicine was designed to redress quantitatively unbalanced humours; degenerate and unnatural *atra bilis* was qualitatively different, and therefore outside the bounds of medical wisdom. Discussions of cancer's origins viewed the mysterious and malign properties of adust melancholy as integrated into the qualities and 'behaviours' of the disease itself, creating a formidable, changeable adversary.

Conclusion

This chapter set out to answer an apparently simple query. What, I asked, did early modern people talk about when they talked about cancer? The firmest conclusion of the chapter is that this is a question worth posing, for we have seen the degree to which the concept of cancer was at once a malleable construction, and a disease of which the fundamental 'character' remained stable even as medical practitioners debated its specifics. Visible throughout early modern sources on the naming, diagnosis and causes of cancer is the urge to turn this disease from a disparate and confusing collection of incidences into a singular and understandable entity. Thus, the often confusing language of cancer consistently returned to a single image, that of a biting creature; the symptoms of the disease were collected into one creature, the crab, and discussions of cause overwhelmingly offered a humoral explanation.

Those unifying urges could only do so much, and anxieties about the un-knowability of this subject consistently resurfaced. Nonetheless, the tone and content of these primary texts has shown that cancer was a disease understood through shaping discourses about its actions and characteristics rather than by the means, now more familiar to us, of a pathology based on its cellular and chemical properties. These discourses would prove influential upon every aspect of early modern conceptualisation and experience of cancer. Belief in humoral causation would affect which therapies were administered for the disease and lead practitioners to look at dietary, environmental and emotional circumstances as they pondered why some people suffered cancers whilst others stayed healthy. Meanwhile, observation of cancer's crab-like characteristics, and speculation about its roots in the 'evil', unclean and gendered substance of melancholy were to play a shaping role in discussions of the disease's nature.

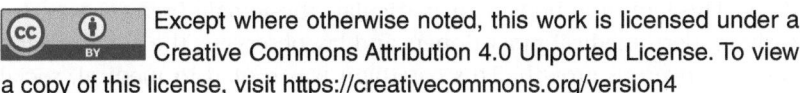

2

Cancer and the Gendered Body

On 3 December 1700, noblewoman Sarah Cowper wrote in her diary: 'My breast is unquiet and gives me troublesome apprehensions. I sometimes seem weary of living, yet find myself often in fear of a painfull lingering death'.[1] Beside the entry was a marginal note in the same hand: 'Fearing a Cancer'. In this chapter, I will argue that Cowper's identification of her breast as the 'troublesome' site where a cancer might breed was, in part, born of contemporary medical and cultural orthodoxy. The feminine body – in particular, the female breast – was, for early modern medical practitioners and lay observers, the paradigmatic site of cancerous growth. This paradigm was rooted in medical, social and aesthetic discourses in which the female body variously appeared as fecund, feeble, dangerous and secret. Moreover, as they attempted to explain cancer's bias toward the supposedly weaker sex, medical practitioners reluctantly engaged with troubling aspects of early modern women's lifecycles, making cancer a disease with the potential to cast light on hidden aspects of the sufferer's conjugal and domestic situation. Women's cancers thus sprang from, and in turn re-inscribed, a model of sexual dimorphism in which the female body appeared physiologically, functionally and pathologically unique.

In exploring the gendering of cancerous disease, this chapter looks in particular to the one-sex/two-sex debate; a discussion which has occupied many scholars since Thomas Laqueur's and Londa Schiebinger's influential proposition of the former model in *Making Sex: The Body and Gender from the Greeks to Freud* and 'Skeletons in the Closet' respectively.[2] In brief, the now well-known 'one-sex' model argues that the notion of two sexes distinguished not only by visible genitalia but by internal pathology was virtually unknown prior to the eighteenth century. Until that point, Schiebinger and Laqueur argue, it was usual to think of woman as an unfinished or imperfect version of man, with her lesser bodily heat causing her to retain inside her body the generative organs which men had on the outside. Thus the ovaries could be seen as equivalent to the testes, and the cervix to the penis. Only in the eighteenth century did other differences – notably, skeletal differences – emerge. This model is largely based on observations of the similitude of male and female genitalia in anatomical texts, and of the popular idea that the ovaries might produce 'seed' similar to that of the testes. From hence, Laqueur in particular posits women's changing social and economic roles as having influenced suppositions about their internal pathology.

Although the 'one-sex' model has proven valuable, several scholars, most notably Michael Stolberg, have persuasively argued that the location of a dimorphic sexual model as emerging in the late seventeenth or eighteenth century is misjudged, and that sexual dimorphism was in fact prominent in texts dating from the sixteenth century onward.[3] As Stolberg points out, 'This is not just a question of getting the dates right: if this is true the contexts from which this earlier discourse of sexual difference emerged also differed from that described by Laqueur and Schiebinger'.[4] His own estimation of possible factors in the development of a 'two-sex' model includes an intellectual shift toward empirical observation as well as commercial gains to be had from medical practitioners' specializing in 'women's problems'.[5] Stolberg's contention is based on a range of evidence, including early modern anatomical drawings and treatises, and writing on sex-specific diseases. In this chapter, I argue that cancer – particularly breast and womb cancers – constituted one such 'sex-specific' disease, which was understood as contingent upon a humoral and anatomical pathology unique to the female sex. It is to be noted, however, that my argument for cancers as linked to sex-specific traits does not preclude a degree of continuity between male and female states. As Gail Kern Paster notes in her 'The Unbearable Coldness of Female Being: Women's Imperfection and the Humoral Economy', the idea that both male and female temperaments could be located on a continuous spectrum, from hot and dry to cold and wet, remained in place even after the notion of genital homology declined.[6] Notably, however, men occupied most of this range. Women, argues Paster, were confined *en masse* to the 'cold and wet' end of the humoral spectrum, with any deviance therefrom taken as abnormal or pathological.

Building upon the theme of 'gendered' illness as confirming sexual dimorphism, this chapter views certain aspects of women's lifestyles as implicated in their physiological and social otherness, and associated susceptibility to cancerous disease. In doing so, I touch upon several aspects of early modern women's physiology and lifestyles for which there are substantial, and growing, critical literatures beyond the scope of this project to examine extensively. Work on menstruation, maternal nursing and domestic violence is notably heterogeneous, with ongoing debate about, for example, whether menstruation was viewed positively or negatively by medical practitioners, whether the use of wet nurses rose or fell over the seventeenth century and how prevalent spousal abuse was in early modern households. In each of these cases, I have dwelt on the points of consensus between authors rather than their differences: that menstruation was a fraught topic, that medical and religious rhetoric favoured maternal nursing and that domestic violence was often permitted within the law.

The first part of this chapter examines the case for viewing cancer as a 'female' disease, showing that although men might suffer from sex-specific cancers, these were rare and not usually attributed to a male pathology. By contrast, women made up the majority of recorded cancer cases, and their sex-specific cancers were believed to be indexed to their distinctly different biology. This sexed biology is the subject of part two, in which I show how the twinned excremental and generative functions of women's reproductive systems were believed to 'breed' cancers. Finally, I consider some environmental factors primarily affecting women and examine why early modern medical practitioners believed that these factors contributed to the development of cancerous disease. Sex, or lack thereof, maternal breastfeeding or refusal to breastfeed, domestic violence and emotional turmoil were all indicated as 'risk factors', such that a woman's cancer might be read as revealing shameful home truths.

2.1   A woman's disease?

In his 'Historical Notes on Breast Cancer', Daniel De Moulin asserts that

> [t]he history of carcinoma was for many centuries mainly the history of breast cancer. Only when in the second half of the 19th century anaesthesia and antisepsis had enabled surgery to treat certain internal carcinomas as well, interest in malignancies other than those of the breast sprang into being.[7]

De Moulin's statement makes some questionable assumptions about early modern surgery, as Chapter 6 will demonstrate. Nonetheless, is it true that in the early modern period, 'breast cancer *was* cancer'?[8] The answer, as this section and this chapter shall demonstrate, is a qualified 'yes'. Breast cancer was certainly the predominant form in most medical accounts, for various cultural, pragmatic and medical reasons. Nonetheless, men did suffer from cancers, as well as being positioned as the 'normal' against which female bodies could be cast as pathological.

It has gone unremarked in the few texts dealing with early modern cancers that, on rare occasions, men were diagnosed as suffering from sex-specific tumours – namely, of the testes ('cods' or 'stones') or penis ('yard'). However, there is some, albeit tentative, evidence for such complaints. A few fleeting mentions of cancers on the yard appear in several medical textbooks around the mid-seventeenth century, usually accompanied by prescriptions for the disease.[9] In the early eighteenth century, John Marten asserted in more detail that 'Swellings or Tumors on the *Stones*', caused by 'Blows, Falls, &c.', could 'terminate into a Cancer' if mishandled.[10] The signs of such a transformation were that 'upon applications to it, it begins to be attended with pricking Pain, &c.', and such cases 'consequently ought not, or but very cautiously to be medled with'.[11] Marten's account relied upon the popular belief, outlined later in this chapter, that bruises could cause cancer. It is notable, however, not for indicating 'male cancers' as a subject area, but rather the opposite; male cancers, even when sex specific, were apparently not viewed as allied to pathological traits peculiar to men, or to gender-specific aspects of their lifestyles. Marten's case appeared in a text dealing primarily with venereal diseases, but it was not implied that cancer should be viewed as just reward for contracting the pox any more than for bruising one's 'cods'. It was simply that this was the circumstance most likely to produce a swelling that could be ill-handled. Moreover, there is no evidence that cancer of the penis or testes was treated, as one might expect under a 'one-sex' model, as equivalent to cancer of the womb.[12] Cures for male cancers appeared either in texts specific to diseases of the reproductive system, or in those dealing especially with cancer, but were seemingly too uncommon to merit mention in the pages of texts on general surgery and physic, where remedies for dermal or breast tumours could be found in abundance.

Overall, only a handful of male-specific cancers were mentioned in early modern medical texts; quite possibly because when it appeared on the genitals, this disease was easily confused with venereal pox, which similarly produced pain, swellings and ulcers, but also because, as I shall contend, theories about the disease's causation meant that medical practitioners did not expect to find cancers here. Neither is there any evidence that when it appeared in men, cancer was thought of as a feminising malady. Conversely, even this unusual 1703 account of a man suffering from breast cancer construed the illness in gender-neutral terms:

> *Hildanus* ... tells of one *Poteer*, an ingenious man, who had a Cancerous Tumour about his left Pap the bigness of a Hens Egg, with which he was troubled many years. Some Physician advised that he would try to dissolve the Tumour and discuss it [with emollients] ... but he no sooner had applyed these to it, but a pain and inflamation arose in the part; so that he was forced to lay that aside and come to the use of a cooling Medicine: The pain and inflamation being allay'd, he applies the Emollients again, but pain succeeded as formerly; and when he found by experience, that these Emollients only raised his pains, and inflamed him, he laid them aside, and the Patient lived a long time after in safety and free from pain.[13]

The subject here is rather the inadvisability of using emollient medicines than *Poteer's* gender, and the patient is approvingly described as 'ingenious'. Another case of male breast cancer can be found in Robert Bayfield's 1655 *Enchiridion Medicum*.[14] Once again, the account is brief and the patient is soon cured with mild medicines. It seems that diagnoses of breast cancer in men during this period were vanishingly rare, and were not linked to gender-specific complaints, as was often the case for women. Where female breast cancer was, as I shall detail, frequently connected to amenorrhea, and hence to the connection between womb and breast, the absence of the womb in men meant that no such conclusions could be drawn. Cases of breast, penile or testicular cancer in men were seemingly viewed as no more nor less allied to their broader humoral makeup than tumours which appeared anywhere else on their bodies.

The contrast between this attitude and that seen in discussion of women's cancers could hardly have been more pronounced. In 1670, the anonymous *An Account of the Causes of Some Particular Rebellious Distempers* declared:

> Cancers are known in part by the Places they fix on, which are the Glands, tho' they may breed in almost all parts of the Body; and this *Aegineta* confirms, who says, a Cancer may happen to sundry Places, as the Lips, Tongue, Cheeks, Womb, and other loose Glandulous Parts; *but were* [sic] *One has a Cancer in any part besides, Twenty have them in their Breasts.*[15]

That view had been orthodox, as Luke Demaitre attests, in the medieval period, and would remain so into the eighteenth century, in which Kaartinen argues that 'having breasts at all was the greatest risk of contracting cancer'.[16] In 1721, for example, *An Universal Etymological English Dictionary* defined 'Cancer' as 'a dangerous Sore, or Ulcer; as in a Womans Breast'.[17] Although it is impossible to determine with any accuracy how many cancers, and what kind, were diagnosed in England between 1580 and 1720, Edward Shorter has found that in parts of eighteenth-century Europe, recorded deaths from cancer were up to nine times higher among women than among men.[18] Furthermore, non-medical texts readily adopted the paradigm of cancer as 'of the (female) breast'. For instance, churchman Thomas Adams's 1615 invective against thieves described them as like 'that disease in the brest, call'd the *Cancer*'.[19] Similarly, in John Webster's 1612 *The White Devil*, Flamineo described himself as 'like a wolf [cancer] in a woman's breast' (5.3.54), while Shakespeare's ambiguous 'canker' often played upon parallels between floral and female bodies.[20]

Cancer was thus paradigmatically a 'woman's disease' in the sense that it was much more frequently identified in women, and that, as both consequence and cause of this bias, the breasts represented the archetypal cancer site. This bias did not mean that men could not suffer from cancers, including some that were sex-specific. However, where men's cancers were generally considered the result of bad diet, bad humours or simply bad luck, women's sex-specific cancers were, as I shall describe, attributed to the peculiar pathology of the female body.

2.2   Breeding a tumour: cancer and female pathology

That women were more likely than men to suffer from cancerous disease was a commonplace in early modern medical and popular understandings of the malady. Exactly why this should be the case, however, remains to be explored, and I contend that women's susceptibility to cancers was explained in terms of their sex-specific pathology, and in particular, their peculiar anatomy. The uterus, the female breasts and the connection between them provided a fertile environment for cancers to grow, flourish and even mimic that most paradigmatically female of bodily states, pregnancy.

Arguably the driver behind all 'feminine' cancers, as well as a host of other female-specific disorders, was one mysterious and much-discussed organ, the womb. Fundamental to generation, and remaining 'secret' within the body, the womb, as Katherine Park asserts, 'appeared as a – arguably *the* – privileged object of dissection in medical images and texts'.[21] Matthew Cobb and Monica Green likewise observe that unlocking the secrets of the female reproductive system seemed for early modern anatomists and medical practitioners a sure route to understanding the mysteries of generation more generally.[22] While they were consistently fascinated by this organ, however, medical texts also reflected cultural ambivalence about the status of the womb, and in particular one of its main functions, menstruation. On one hand, it was widely accepted that, as Stolberg points out, menstruation provided a system by which excess humours, gathered in the womb, could be expelled from the body, thus preventing illness.[23] Haemorrhoidal bleeding in men was commonly viewed as an imitation of that process, as were periodic nosebleeds.[24] On the other hand, however, most medical practitioners believed that women only required such a system because of the lack of perfecting heat in their bodies, which was inadequate for the full concoction or perfection of the blood.[25] In Stolberg's words, '[T]he need for menstruation, not the evacuation itself, was pathological'.[26]

While menstruation might be a healthy process, menstrual blood was sometimes – particularly prior to the seventeenth century – viewed as excremental and noxious, to the point that certain medical writers believed the proximity of a menstruating woman could kill plants, sour milk and cause infants to become sick.[27] Furthermore, throughout the early modern period, the womb was commonly viewed as an unreliable organ, prone to dysfunctions which threatened not only the woman, but her unborn children, her family and society at large. The terms in which these dysfunctions were presented were often lurid, explicitly depicting the womb as a negative, though necessary, constituent of the feminine body, which was partly independent of the woman in whom it 'resided'. In 1636, for example, John Sadler wrote in *The Sick Woman's Private Looking-Glasse* – purportedly aimed at a female audience – that 'from the wombe comes convulsions, epilepsies, apoplexies, palseyes, hecticke fevers, dropsies, malignant ulcers, and to be short, there is no disease so ill but may procede from the evill quality of it'.[28] Still more dramatically, a translated work by the French physician Jean Riolan, printed in 1657, insisted that

[t]he womb is the Root, Seed plot and foundation of very near al womens Diseases, being either bred in the womb, or occasioned thereby.

If it be troubled with an hot distemper and inflamed, it causes intollerable burnings, the Feaver Synochos and the burning Feaver, very troublesome Itchings and finally it brings exulcerations, the Cancer and Gangraena.

If it be stung with fervent Lust, it becomes enraged, causes Uterine fury and Madness; wil not let the Patients rest, but invites them to shake and agitate their Loins, that they may be disburthened of their Seed; and at last, they become shameles and ask men to lie with them.

Somtime it is drawn out of its place towards the sides, and is carryed this way and that way, as far as the Ligaments and Connexions of the Womb wil give leave; and it wil rise directly to the Liver, Stomach and Midrif, that it may be moistened and fanned; it Causes Choaking and Stranglings, and raises terrible and violent motions and Convulsions in the Body.

In a word, the Womb is a furious Live-wight in a Live-wight; punnishing Poor women with many Sorrows.[29]

In this description, the womb acted in ways which made clear that it had no functional counterpart in the male body, threatening the life of the afflicted woman, and disrupting familial and societal structures by inducing inappropriate lust. It was, like cancer, both of and hostile to oneself, 'an Animal in an Animal', imbued with a degree of sentience and, according to some, 'Brutish understanding'.[30] Accordingly, one common remedy for the 'Mother', or wandering womb, was to tempt the organ back into its proper place by holding foul smells at the nose and sweet ones under one's skirts. Some sources even attested that the womb continued living for some time after a woman's death.[31]

As Riolan noted, the temperamental womb was also susceptible to cancers. Indeed, it was the only internal organ for which diagnoses of cancer were consistently, if not frequently, advanced. As we have seen, cancers of the fundament or intestines appeared only very occasionally in medical texts. Cancers of the womb, however, were described in more detail in a number of writings across the early modern period, in terms which reiterated medical ambivalence toward that organ. The important visual symptoms of cancer, described in Chapter 1, were obviously absent from these diagnoses and replaced by sensational ones, including pain, amenorrhea, difficulty in urinating, feelings of heaviness and tiredness.[32] Somewhat problematically, such symptoms were common to many renal and gynaecological conditions, not least pregnancy. To clarify the situation, Lazarius Riverius suggested that one might use 'a Womb-perspective Instrument' to locate the problem.[33]

Medical practitioners might also manually examine patients in whom they suspected uterine cancers. For example, the physician and surgeon Edmund King wrote in his casebook that examining a 'Mrs Hutchinson', who complained of constipation and pain in her groin and abdomen, he had 'felt in vagina...noe passage bigger than to admit the end of a little finger or swan quill'.[34] His tentative diagnosis of a tumour in the *'cervix uteri'*, however, was only confirmed by Hutchinson's death and post-mortem.[35]

Riverius's 'Womb-perspective instrument' never took off, and manual examinations such as King's were rarely conducted (or, perhaps, rarely recorded). In the absence of reliable means of internal examination, the surest sign of an ulcerated cancer in the womb, agreed upon in most medical texts dealing with this subject, was a foul 'sanies', or discharge. Medical practitioners dwelt at length upon this symptom. Robert Bayfield, for instance, talked of a 'carrion-like filth' in the womb, while Paré asserted that the disease 'poures forth filth or matter exceeding stinking & carion-like, and that in great plenty'.[36] Others described the womb as issuing 'a blacke graene matter', which was 'cadaverous'.[37] The emphasis on these substances as unclean was more concentrated than anywhere else in discussions of cancer – it was the definitive sign of the disease, rather than an unfortunate side-effect. Descriptions of 'filth' emanating from the womb clearly echoed fears about the potentially harmful properties of menstrual blood. In the positioning of such matter as 'carrion-like' or 'cadaverous', writers also raised the disturbing image of a disease consuming the body from the interior, just like a rosebud eaten from within by a canker.[38]

Given contemporary ideas about the humoral causes of cancer, the womb's supposed susceptibility to this disease, and the language in which its symptoms were described, are unsurprising. The womb provided a sink for what Riverius described as a 'perpetual Common-shore of Excrements': humours which were viewed as, at best, surplus to requirements, and at worst, degraded and feculent.[39] When not expelled through the menses, these humours could accrue and stagnate in precisely the way believed to breed tumours. As such, restoring menstruation which had stopped unexpectedly was described as a matter of urgency in texts dealing with all kinds of cancer in women.[40] The reasons for amenorrhea were diverse, and, as described elsewhere in this book, sometimes environmental. One obvious factor, however, was age. Though it was not generally emphasised, medical practitioners could not help but observe that 'Of twenty Women afflicted with Cancers, fifteen will be found to be aged from forty five to fifty Years, when

Nature usually puts a stop to the menstrual Evacuations'.[41] Diagnoses of cancer in menopausal women inevitably intersected with prevailing medical and cultural discourses which Stolberg argues positioned the menopausal woman as weak and in precarious health by dint of her cooling humours.[42]

Another obvious means by which the menses might be suddenly interrupted was pregnancy. Although there is no evidence of confusion between the two conditions, it is notable that many of the initial symptoms of conception were cruelly mimicked by uterine cancer. Indeed, 'moles', or false pregnancies – identified by some onlookers as the cause of Mary Tudor's false conception in 1554 – were believed to be masses of tissue somewhat akin to tumours, though, crucially, lacking the malignancy characteristic of cancers.[43] More broadly, it is evident that, following on from the attribution of zoomorphic sentience to cancers, the disease – in the womb, but also elsewhere – could be perceived as a variety of 'monstrous progeny'. Chapter 4 discusses medical practitioners' habit of comparing cancerous tumours at every stage with organic objects with marked potential for growth or generation, such as seeds, nuts and eggs. Cancers were also repeatedly characterised as having been 'bred' from ill humours, and contemporary interest in spontaneous generation, as described in Chapter 3, vivified the long-held belief that tumours might contain 'al kynd of humours, but also sound bodies, and straunge thinges'.[44] Most strikingly, throbbing pain in a tumour was sometimes characterised as pulsation.[45] In 1583, for instance, Philip Barrough asserted that '[a]bout the place where cancre is lodged, there is felt a certaine beating or pulse, and as it were a pricking: sometime also (as Celsus saith) the tumour is a sleepe, and as it were deade'.[46] In this context, a cancer's 'breaking out' from the body might be viewed as a grotesque delivery which imitated the dangers of childbirth.

In the case of cancer, the ambivalence traditionally present around the womb was thus particularly strong. Both the excremental and generative functions of the womb fitted with perceptions of how cancerous tumours came about, and the womb's quasi-independence from – even hostility toward – the body in which it 'resided' echoed that attributed to cancer. Nonetheless, womb cancers were recorded only rarely compared to tumours in the breast. The reasons for this apparent contradiction inhered in the supposed peculiarities of female biology and the practicalities of diagnosis. As Chapter 1 describes, medical practitioners noted the near impossibility of diagnosing internal cancers. Even the 'sanies' which might accompany uterine cancers were an uncertain sign, and patients may have been reluctant to consult upon (and doctors reluctant to record) a symptom which was also characteristic of some varieties of venereal pox. In any case, it was generally accepted that, while they might be palliated, there was no effective cure, pharmaceutical or surgical, for such complaints. For the early modern medical practitioner, however, disorder in the womb did not necessarily mean that a cancer would arise in that organ. Other, more easily diagnosed, spots could bear the brunt of excremental humours, and first among these was the vulnerable and desirable female breast.

According to most early modern medical textbooks, the womb was, by one means or another, connected to the breast, more directly than to any other part of the body.[47] For many writers, the connection was a simple physical one, outlined in the seminal works of Galen and Hippocrates and confirmed by their own investigations.[48] In 1657, for example, Riolan asserted confidently that

> There is a great League, and fellow-feeling, between the Dugs, and the Womb, by reason of two Veins, *viz*. The *Vena Mammaria,* or Dug-Vein; and the *Epigastrica:* and also by the *Venae Thoracicae,* or Breast-Veins, which are Branches of the *Vena Cava,* which in the bottom of the Belly, affords the Hypogastrick Vein unto the Womb.[49]

Other practitioners supposed a different arrangement of connecting vessels, or a vaguer 'consent' between the two organs, but it was commonly agreed that the two 'communicated'.[50] As the anonymous *An Account* observed, '[T]he Breasts of Women are tender...which upon the flowing of the Courses, that tenderness leaves them'.[51] Further evidence could be found in the way that post-partum women did not menstruate, but did lactate. According to many eminent practitioners, blood which was usually surplus, and hence excreted as menses, was used during pregnancy to sustain the foetus, and was afterwards diverted to the breasts to make milk.[52] Breast milk might thus be viewed as 'nothing but the menstruous bloud made white in the breasts', having been altered by divine design in order to avoid the alarming sight of infants covered in blood.[53] Under this model, the female breast was functionally unique; rare reports of male lactation merely imitated the same process.

For those writers concerned with cancer, it was apparent that the connection between breast and womb could endanger as well as sustain life. If nutritive blood might travel from womb to breast in order to be concocted into milk, it was also possible that the excremental, possibly harmful humours associated with menstrual blood could make the same passage. *An Account* further explained that 'The Ancients observ'd, that Women were most troubled with Cancers, upon the stopping of their Monthly Visits', because when bad humours were not discharged through the menses, they were most likely to 'discharge themselves' on the breasts.[54] That conclusion was shared by medical practitioners across the early modern period, though exactly what was transported, and by what mechanism, was a matter for debate: was it melancholy, *atra bilis* or another kind of 'burnt Blood'?[55] Some medical practitioners seemingly believed that the connective structures themselves could also become diseased, though this view was uncommon: John Ward, for example, recalled in his diary a conversation with Walter Needham, in which the eminent physician informed him that in one post-mortem examination 'hee hath seen a string...going from the breast to the uterus. I suppose it was the mammilarie veins full of knots which were cancrous, and hung much like ropes of onions'.[56]

Furthermore, breasts were not only rendered vulnerable to humoral 'discharge' by dint of their direct connection to the womb. Rather, susceptibility to absorbing excess humours was a characteristic of the breast itself – or more accurately, the female breast, since the flesh thereof was widely accepted to be of a 'Glandulous' quality. According to the 1656 *The Compleat Doctoress*, 'The Breasts are naturally thin, spongy, or funguous, and loose; for this reason they are apt to entertaine any crude and melancholy humours, flowing to them either from the Matrix, or from any other parts'.[57] The female breasts' lax structure could be evidenced by palpation and anatomical examination. They were, in most cases, and especially in the older women most susceptible to cancers, visibly larger and less muscular than the male equivalent, differences which were not only visually but medically significant. Moreover, discussions of these tissues' laxity often bore a misogynistic taint. Large breasts, it was suggested, provided a particular abundance of 'loose' flesh in which to breed a cancer:

> [T]he swelled Breaths of Ancient Virgins and married women, are liable to the same Diseases. For either by reason of a Flux of Humors or of some bruise, they are inflamed and impostumate...Hence comes an incurable Cancer; Because the Dugs are ful of Kernels and spungy, and therefore ordained by Nature to receive superfluous Humors.[58]

The fleshiness which allowed 'superfluous' humours to gather and form tumours was, for this 1657 text, directly indexed to two kinds of women with minimal libidinal capital, old maids and wives. Elsewhere, large breasts were deemed both 'very unsightly', and indicative of lustfulness, such that, as Paster contends, '[t]he large breast is the female metonymy not only of age but of shame and thus of a specifically gendered form of social and bodily inferiority'.[59] As *The Compleat Doctoress*'s observation of the breasts 'entertaining' crude incoming humours suggests, loose and lax breasts were often thought to indicate loose and lax women, since many believed that 'the cause of [the breasts'] greatnesse is often handling of them' or 'stroaking of them'.[60] The popular *Compleat Midwife's Practice*, meanwhile, linked breast size and its associated dangers to greed, when it advised that women alter their diet to reduce the breasts, since 'the lesser the Breasts be, the less subject they are to be cancered'.[61] Once again, these bodily responses were at least partly sex-specific. Women's inability to resist either gastronomic or sexual temptation could be ascribed to their naturally weak characters, in contrast to the self-mastery supposedly exercised by men.[62] In addition, it was believed that older women in particular had 'colder', sedentary bodies in which fat was more apt to congeal and less likely to be fully 'concocted' into blood and spirits. In a literal sense, the female body burned fewer calories.[63]

Medical explanations for the prevalence of breast cancer diagnoses over all other types thus engaged with wider cultural ambivalence about female breasts more generally. It is clear that breasts were sites of sexual desire, both for men looking upon them, and according to Riolan, for women too. 'In ripe Virgins fully Marrigable', he asserted, 'the Dugs are firm and solid':

> They become more soft and swelling, when they are transported with a burning desire of carnal Embracements: and by how much the higher they swel without pain, and the fuller Orbe that they make, strowing and Kising one another, the greater is their desire after bodily Pleasure, and it may be guessed that they have tasted the Sweetness of Mans-Flesh.[64]

Writing on the significance of these 'orbs', scholars including Angela McShane Jones and Gail Kern Paster have noted the trend for exposed breasts in fashionable dress during parts of the seventeenth century.[65] Looking to art, fashion and literature, Marilyn Yalom similarly contends that '[t]he meaning of the breast in Renaissance high culture was unequivocally erotic'.[66] Exposed breasts could signal fecundity and erotic potential. Furthermore, the nipples of the breasts were occasionally compared to the head of the penis, 'in that by handling or sucking it becomes erect or stiff'.[67] In related discourses, women were occasionally described as 'milking' the penis during sex, whilst breast milk was itself a remedy for male impotence.[68] Viewing the breasts in these terms did not preclude writers or artists from also valorising their maternal function, and noble-women were sometimes painted bare-breasted, surrounded by their children.[69] However, such positive representations of the breasts were strictly conditional; as Margaret R. Miles observes, breasts were commonly represented as either 'extremely perfect' or 'extremely bad'.[70] To be extremely perfect, the breasts, and the individual to whom they belonged, needed to fulfil a raft of criteria. The breasts should be small, high and youthful, promising the fertility of the bearer; furthermore, she should be modest, chaste and of aristocratic pedigree, as well as (preferably) available for marriage. Breasts which became cancerous might have, by dint of their size and age, failed the demands of perfection even prior to illness. When they became diseased, they offered a sign of illness and decay which was in stark contrast to the erotic and maternal ideals of youth, fecundity and plenitude.

The status of the female body, and more specifically, the female breast, as a paradigmatic site of cancers in this period thus depended on discourses in which ambivalence and mistrust toward sex-specific organs was long established. On one hand, the womb and the breast both possessed the mysterious power to nurture and sustain life. On the other, medical practitioners widely accepted that such generative power was bound up with women's constitutional inability to perfect the matter of their humours, and therefore the contingency of their health on menstruation. As women approached older age, this paradox became increasingly fraught, and the womb appeared, like cancer, as both of and hostile to the body, moving around uncontrollably, and creating monstrous growth. That these concerns were transposed onto the breast reflects both contemporary beliefs about the porosity of that organ, and the pragmatic limitations of early modern diagnosis. The womb was impossible to view in a living patient and produced unreliable symptoms. The breast, however, provided a visible, palpable site from which the destructive and constructive potential of the uterus could be read.

2.3   Domestic bodies: cancer and female lifestyles

Women were viewed as uniquely vulnerable to cancer, and in particular to breast cancer, for a number of biological reasons. Yet, early modern practitioners noted the obvious: not all women, menopausal or otherwise, suffered from the disease. As detailed elsewhere in this book, several non-gendered factors were believed to influence one's susceptibility to cancer, how fast it progressed and if it might be cured. However, many of the elements medical practitioners identified as rendering one at risk of the disease were, implicitly or explicitly, those which linked the peculiar physiology of women to social or domestic phenomena which were either sex-specific, or affected women to a greater extent than men. This section looks at several of the most prominent: maternal nursing, sex, domestic violence and emotional trauma.

Demonstrating the indivisibility of social and biological bodily functions in the early modern period, the most widely discussed 'risk factor' in texts about cancer, as well as discussions of that disease in household receipt books, midwifery texts and manuals of physic, was the thorny issue of maternal breastfeeding. Lactation, as described earlier, was often thought to involve the flowing of humours into the breasts for concoction into milk; a process which, in contrast to the noxious 'discharge' of excremental humours into that tissue, was essentially healthy. As was often the case in discussions of cancer's cause, however, medical practitioners feared that this healthy process might, for a number of reasons, turn unhealthy. Prone to inflammatory infections such as mastitis, the lactating breast was viewed as a potentially vulnerable organ. In particular, medical practitioners knew that problems arose when, for whatever reason, milk stayed in the breasts and stagnated there. For an early modern audience consistently exposed to religious, cultural and medical debate about the advisability of maternal nursing, that fact was particularly important. As Valerie Fildes and David Harley have documented, the seventeenth and eighteenth centuries saw a steady rise in the number of medical practitioners touting maternal nursing as preferable to wet-nursing, though not necessarily a corresponding shift in behaviour.[71] The 'failure' of upper-class women to nurse their own infants was, argues Harley, increasingly cast as an issue of public moral and physical health, and, then as now, women who 'refused' to breastfeed were often cast in lurid terms. One 1612 work on childbirth, for example, asserted that there was 'no difference betweene a woman that refuses to nurse her owne childe; and one that kills her child, as soone as she hath conceived'.[72]

The increased risk of breast cancer attendant upon failing to breastfeed one's children was explicitly stated in several medical advice books, from across the early modern period, which held that milk became dangerous when it 'curdled' or 'coagulated' in the breasts.[73] In 1671, midwife Jane Sharp stated that

> [i]f there be too much milk in the breasts after the child is born, and the child will not be able to suck it all, the breasts will very frequently

inflame, or imposthumes breed in them; they swell and grow red, and are painful, being overstretched, where hard tumours grow: too much blood is the cause of it, or the child is too weak, and cannot draw it forth.[74]

These unspecified 'imposthumes' could easily turn cancerous. Notably, however, such texts did not argue for the immorality of the non-nursing mother, nor cast cancer as her 'punishment'.[75] Rather, they made conspicuous efforts to explain why one might not nurse, or nurse inadequately, and suggested alternative means for drawing milk from the breasts, including suckling by puppies, by another woman or by 'an instrument designed for that purpose'.[76] Medical practitioners' apparent disinterest in blaming a non-nursing mother for her cancer was born of several factors. There was, as shall be seen later in this chapter and in the book, a general disinclination to assign blame for cancers. People with cancer were acknowledged to be suffering immensely and usually mortally, and attracted much sympathy. They were also, in the eyes of medical professionals, valuable paying customers. In addition, though they commonly agreed that breast cancer and lactation were linked, medical practitioners were often cagey about whether breast-feeding actually diminished or increased the risks of cancer. Shorter's *A History of Women's Bodies* records that, in 1798, one continental doctor complained that a 'folkloric belief that lactation caused breast cancer' was responsible for women's refusal to breastfeed.[77] That 'folklore' may well have been contemporary wisdom in the seventeenth century, when one anonymous household receipt book grouped together cancers of the breast with 'nipping biting in the breasts by giving Children suck'.[78] Several more medical writers acknowledged a connection between lactation and breast cancer, but were vague as to whether the risk was exacerbated by breastfeeding.[79] The early modern woman thus faced something of a double bind in relation to this 'risk' factor. Lactation, it was acknowledged, increased personal susceptibility to cancer, but how mothers might sidestep this physiological hazard by altering their behaviour was uncertain, and would remain so for decades to come.

Where lactation presented a biologically unavoidable risk to new mothers, the social structures which made motherhood more generally a woman's duty were also implicated in cancer's cause, often in contradictory ways. Marriage and childbearing almost always represented the most proper and 'natural' lifestyle for an early modern woman.[80] Texts on cancer sought neither to diminish nor support this institution, but showed how marriage, spinsterhood and celibacy all presented biological hazards. It was repeatedly (though still infrequently) observed during this period that nuns appeared particularly susceptible to breast cancer. Dionis, for instance, observed in 1710 that 'the Disease is very rife in Nunneries'.[81] Meanwhile, Madame de Motteville remembered her mistress, Anne of Austria (Queen Consort of France and later regent for her son, Louis XIV), as having on several occasions visited nuns 'all rotten' with breast cancer, recording on one occasion in 1647 that '[t]he disease had so eaten away into the part on which it had fastened that we could see into [the nun's] body'.[82] This link between nuns and cancer seems to have prevailed for much of the early modern period, and across national borders.[83] Investigating incidences of breast cancer in Italian and Spanish nunneries, Sarah E. Owens cites the Paduan medical practitioner Barnardino Ramazzini, who attested in 1713 that 'tumors of this sort are found in nuns more often than in any other women... Every city in Italy has several religious communities of nuns, and you seldom can find a convent that does not harbor this accursed pest within its walls'.[84] Cancer was in these instances understood as resulting from a combination of sex-specific physiological and circumstantial factors. Simply put, lack of sex meant that a woman had no opportunity to put her 'seed' to use in the creation or nourishing of a child. To expel the seed (concocted blood), nuns therefore needed to menstruate more, and if they did not, they would likely suffer with one of the many diseases caused by excess humours either collecting in and blocking up a part of the body, such as the circulatory vessels of the breast, or stagnating and putrefying in the womb, from whence noxious vapours could affect the stomach and brain.[85]

Celibacy, enforced or elective, thus presented a serious risk to women's health. However, writings on cancer also made clear that married life – the only acceptable sphere for female sexual activity – held its own dangers. Throughout sixteenth-, seventeenth- and eighteenth-century medical texts, the tendency of cancer to follow a bruise or fall was prominent.[86] Multiple medical textbooks suggested that 'blows, strokes, punches', 'falls or bruises', 'a Blow, or some Bruise' or 'a fall, a stripe, a blow, a bruise' were among the most likely causes of cancer, particularly breast cancer.[87] The physiological basis for this statement was clear. Anyone looking upon a bruise could see the discoloured blood welling under the skin, and conclude that the blue, green or yellow tinge thereof represented a stagnation of melancholy and choleric humours in the part, precisely the substances believed to provoke cancers. The perceived causal link between bruises and cancer was so well established that in 1729, a man was brought to court, though acquitted, for causing cancer by punching a woman in the breast on the street.[88] Most strikingly, in 1670, *An Account* gave numerous examples of cancer patients whose tumours appeared after a violent experience:

> we have instances without number, of Women that have had them [cancerous tumours] by Blows, Bruises, &c. as before we have made mention of; and as was the case of a Gentlewoman, whose Husband after a Drunken Bout was thrown into a Fever, and being delirious, upon her giving him something to drink, he hit her Left Breast with his Hand, which caus'd it to Cancerate, of which she soon after dy'd.[89]

> A poor Working-Woman, by a Blow upon her Right Breast with the Key of a Door, which she run against, had a great Pain in it that she could not Rest Night nor Day; the Bruise inflam'd and Swell'd, she ran from one to another for help, till at length she was told it was a Cancer, and must be cut off.[90]

> A Gentlewoman by a punch upon her Breast by a Man's rushing by her in the Street, had such a Pain, throbbing, and at length Inflammation and Swelling, that she was told it was a Cancer [91]

Each of these cases was individually plausible and reinforced the connection between bruising and cancer. Notably, they all involved the breast, suggesting that the damage caused by a bruise was exacerbated by that organ's 'natural' tendency to receive and absorb excess humours. Taken as a body, however, the unusual detail supplied in these stories becomes conspicuous. The gentlewoman received a blow because her husband was not only drunk, but delirious and feverish; the working woman was hurt by the key of a door *which she ran into*. Overall, one feels that, as Porter has observed of grotesque bodies, 'the disclaimer doubles as an attention-seizing strategy'.[92] These accounts actually make more visible the most likely way in which a woman could sustain 'a fall, a stripe, a blow': domestic violence.[93]

The prevalence of spousal violence during the early modern period has been discussed at length by, among others, Garthine Walker, Elizabeth Foyster and Laura Gowing.[94] Though they emphasise different aspects of the wide variety of activities one might characterise as abusive, they all make clear that early modern married women had relatively little legal protection from husbands who might mentally and physically subjugate them, including as a mode of 'reasonable correction'. Women had no right to a separation unless the violence inflicted upon them was deemed life-threatening, and thus might find themselves in situations which imperilled their physical and mental health without legal, economic or practical means of escape.[95] Not all domestic violence was spousal, and women were also known to enact violence upon servants, children and spouses. Nonetheless, male-on-female violence appears to have been more common, and seems implicit to *An Account*'s convoluted tales of how three women found themselves receiving blows to the chest which had nothing whatsoever to do with the dispositions of their husbands, fathers or masters. Medical practitioners' reluctance to identify domestic violence specifically as a cancer cause is understandable, since to do so would cast aspersions on the situations of those whom they treated for the disease, not to mention their spouses.[96] Writing in her diary, however, the formidable gentlewoman Sarah Cowper experienced no such compunction. On 23 February 1700, she wrote, with character-istic candour, that '[a] visitor told me it was said the Lady Ang. was like to dy of an Ulcer in her Womb and a Cancer in her Breast both caused by the Barbarous Cruelty of her L[ord] ... with the utmost detestation [I] cou'd see scourged this cruel, brutish L[ord]'.[97]

Cowper's assessment of 'Lady Ang.'s ill health, clearly passed on by a gossiping acquaintance, shows the popular currency of the 'bruise' theory of cancer causation. It also shows how, outside medical text-books, the physical effects of violence could not be separated from its emotional and social ramifications. Medical practitioners identified grief, anger, brooding and mourning as possibly contributing to the develop-ment of cancers in both sexes.[98] Women, however, were once again at particular risk from a combination of physiology and personal circum-stances. Even in normal, peaceful settings, women were thought to be constitutionally less able to moderate their emotions. Evelyne Berriot-Salvadore summarises: 'According to a tradition stemming from Aristotle and others, woman was weak, quick to anger, jealous, and false, whereas man was courageous, judicious, deliberate, and efficient'.[99] Being on the receiving end of domestic abuse (emotional or physical) thus necessarily had a particularly strong and uncontrollable effect on the female sex. In women's accounts of violent marriages, fear, as one might expect, featured strongly.[100] One had to be in fear of one's life in order to justify a court separation, and such an extreme of emotion might be expected to have a damaging effect on already fragile female constitutions. A husband did not necessarily have to beat his wife, however, in order to bring about grief, anger, sadness and potential physical harm. Only months after recording the 'Barbarous Cruelty' of Lady Ang.'s husband, Cowper wrote:

> A lady of my acquaintance had a Cancer broke in her Breast…it was thought the result of a foul disease she got of her Hus[band], who was known to be a Proffligate man. These are sore calamity, but what gives them inexpressible weight is that (perhaps to palliate his own crimes), he accused her of a design (confederate with the Butler, I think it not likely) to poison him.[101]

Her account bespeaks a complete breakdown of the conjugal relationship, a story of betrayal, recrimination and counter-accusation. Transmission of venereal diseases was, as Gowing notes, sometimes cited as a manifestation of 'cruelty' in separation cases, since it caused physical damage.[102] Moreover, cancer in this case became, while not a 'shameful' disease as such, a means by which the unsavoury and potentially shameful details of one's domestic circumstances could be surmised by others. Sources such as Cowper's diary are rare, but her entries suggest that some onlookers, medical or otherwise, might have heard of a woman's cancer and begun to speculate about her life behind closed doors.

Conclusion

In Hephizibah Roskelly's 2012 account of her experience of breast cancer, she dwells upon the seeming betrayal of the mind by the body. 'My feminist thinking', recalls Roskelly, 'had to be rethought when I got the word that something toxic – potentially fatal – lived inside me, and had for awhile, long enough that a body that was nurturing the mind…could have mentioned something'.[103] Though cancer may no longer formally be considered a 'woman's disease', discourses of risk and debates over treatment remain congregated around the female body, and many of the hot topics in these debates – breastfeeding, childbearing, the effect of grief – remain strikingly similar to those I have identified for the early modern period. Moreover, these accounts seem in places to prefigure Roskelly's sense of the female body as having its own agenda, potentially acting against the person 'inside'. In early modern medicine and culture, it was often accepted that women's lives must be blighted by ill health. Because of their unstable humours, their emotional incontinence and their 'destiny' to bear children, women suffered from an array of sex-specific diseases. Textbooks discussing women's health issues far outstripped similar texts about men and underlined this sex's status as not only fairer but weaker. The gendering of cancer as a disease to be found primarily in the female breast was largely a product of this discourse, trading on speculation about women's mysterious anatomy and in particular the 'secret' womb. Cancer texts also recognised that women's lifestyles presented several 'risk factors'. Mindful of their market, medical practitioners were reluctant to state in print that domestic turmoil, and choices (or lack thereof) around breastfeeding and sexual activity, might predispose one to cancerous tumours. Free from such concerns, however, Cowper's diary, providing a fascinating and rare glimpse into lay perceptions of cancer, shows that readers might be all too aware of what medical texts really meant when they described the risks of 'grief' or 'blows', and from whence the greatest risk of these arose for women – their marriages, their masters or their parents.

Cancer might also be viewed as representing the 'pathological' nature of women's bodies more fully than other diseases. Cancerous tumours were both a part of the body, generated and sustained by the humours, and a hostile interloper, eating up one's substance. This paradox closely matched that understood to characterise women's peculiar physiology. The bodily phenomena which made women able to bear children – the womb, the 'coldness' of the body and the excess of humours to be voided through menstruation – were the same as those which 'betrayed' them and so frequently made them ill. More broadly, the generative function was a hazardous one in its own right, since childbirth represented the most perilous event of an early modern woman's life. In constructing cancer as a 'gendered' disease, early modern writers thus depicted the illness as both contingent upon, and imitative of, the double bind of women's life-giving but dangerous bodies.

3

'It Is, Say Some, of a Ravenous Nature': Zoomorphic Images of Cancer

In Chapter 1, I described the crab as the oldest and most pervasive zoomorphic image of cancer, bound up with the disease's etymology and diagnosis. This creature, however, was arguably the least colourful, and certainly the least frightening, of several animals which came to be associated with cancerous disease. In this chapter, I shall argue that the most extreme and culturally resonant figurations of cancer during the early modern period were to be found in the unlikely pair of the worm and the wolf. Through examining the use of these beasts as both popular and medical images, I discuss why early modern Englishmen and women came to associate these creatures with cancer, and how the cultural freight of worms and wolves shaped, and was shaped by, anxieties surrounding this disease.

The relationship between human and non-human species in the early modern period has proven a productive field for literary and historical scholars of the past decade, though it remains under-explored within the medical humanities. Studies of the human/animal interface have often focussed on the anxieties generated by incomplete or fragile distinctions between (wo)man and beast, and on creatures which seemed to bridge the gap between the two. Taking its departure from Keith Thomas's influential *Man and the Natural World*, Erica Fudge, Ruth Gilbert and Susan Wiseman's edited volume *At the Borders of the Human* offers a collection of essays considering bestiality in humans and humanity in animals, of which Margaret Healy's 'Bodily Regimen and Fear of the Beast' has a particular influence on this chapter.[1] More recently, Jean E. Feerick and Vin Nardizzi's edited collection titled *The Indistinct Human in Renaissance Literature* has sought to expand upon the topic by offering essays which dwell upon the animal, vegetable and mineral contexts of Renaissance experience, seeking an ecocritical decentring of the human subject.[2] Ian

61

MacInnes's contribution to that volume, 'The Politic Worm', provides the most comprehensive analysis of invertebrates in Renaissance culture to date and is discussed further in the latter half of this chapter.[3] It is notable, however, that despite focussing closely on the worm in the human body, MacInnes does not mention the 'worm' of cancer or its relation to the horticultural canker-worm, an omission perhaps owing to current lack of scholarship on cancers in this period.

Elsewhere, scholarship on individual texts or authors has also provided insight into the rhetorical uses of animals in early modern culture, often centring on religious works. Karen Edwards's 'Milton's Reformed Animals' provides a comprehensive collation of the occurrence and significance of animals in that poet's work, which informs various parts of this chapter.[4] Marta Powell Harley and Jonathan Wright have looked to the worm to shed light on Chaucer's 'Physician's Tale' and Reformation religious tracts, respectively.[5] Most significantly for this chapter, Jonathan Gil Harris's analysis of the utility of the canker-worm in Gerard Malynes's *A Treatise of the Canker of Englands Common Wealth* is the only literary-focused work to draw the connection between canker-worms and cancer, usefully arguing that the former lent a 'distinct, ontological agency' to the latter.[6] As will become clear, however, I believe that the connection Harris portrays might benefit from closer attention to the materiality of the cancer-worm.

Drawing from this rich critical field, this chapter focuses on two creatures consistently and often problematically associated with cancerous disease in the early modern period. My first section examines the wolf, a creature long associated with cancers because of its ravenous, secretive nature. The second, longer, section of the chapter considers the worm and explores the linguistic and scientific basis of 'cancer-worms' and their significant cultural freight.

3.1   The wolf

> [Thieves] lye in the bosome of the *Church;* as that disease in the brest, call'd the *Cancer,* vulgarly the *wolfe:* devouring our very flesh, if we will not pacifie and satisfie them with our substance.[7]

In 1615, clergyman Thomas Adams chose the twinned images of wolf and cancer to express his loathing for those who stole from the church, in a collection of three sermons titled *The Blacke Devil or the Apostate, Together with the Wolfe Worrying the Lambes, and the Spiritual Navigator, Bound for the Holy Land.* Adams's designation of cancer as a 'wolfe' pointed to anxieties about the destructive potential of certain godless individuals within the body of the Church. It depended on ideas about wolves formed in religious discourses, many of which spilled over into dramatic and poetic forms of writing. Moreover, the sermon recognized and reiterated the long-standing association of cancer and wolves, in which medical practitioners and popular writers variously compared cancer with a wolf, used 'wolf' as an alternative name for cancer or even believed the disease to be literally a wolf in the body. The variety of ways in which the wolf emerged as a 'cancer animal' reflected the range of beliefs which might arise from one potent central premise: that being devoured by an animal was an appropriate metaphor for the degeneration effected by a malignant disease.

To examine these discourses, I shall begin at the most extreme end of the spectrum of beliefs about the cancer-wolf. Here, one finds an extraordinary, and unusual, account from the respected physician Daniel Turner.[8] Turner noted that cancer, being a disease difficult to cure, attracted many tall tales about its nature and causes. Such a tale, he wrote,

> I was not long since inform'd of, by a Woman who vow'd, that in Time of Dressing, one of these Ulcers, by a villainous Empiric (a famous Cancer Doctor) when they held a Piece of raw Flesh at a Distance from the Sore, the Wolf peeps out, discovering his Head, and gaping to receive it.[9]

Turner's anecdote may seem unbelievable. Yet underlying the story of the 'villainous Empiric' and his patient was a web of convictions about the nature of cancerous disease which in their most extreme form could lead to belief in the 'wolf' of cancer as a bodily reality. Foremost among these beliefs was the observation that cancers seemed to 'devour' the body, growing larger as the patient became steadily more emaciated. This belief was fostered in part by widespread attestation of the efficacy of 'meat cures' such as Turner described; that is, the palliative application of freshly killed and sliced poultry, veal, kittens or puppies to a cancerous ulcer. By offering the devouring cancer a meal that was warm, fresh and appealing, it was believed, the disease could be tempted to stop eating the patient, at least for a time, and consume the meat instead.[10]

Faith in the meat cure did not necessarily imply that one believed, like Turner's empiric, that a wolf could literally be present in the human body. Nonetheless, the therapy sprang from, and reinscribed, an image of cancer as flesh-eating which made stories such as this one imaginatively satisfying. Meat cures were widely used, and the connection between this therapy and the cancer-wolf was long established. In the fourteenth century, for example, surgeon Guy de Chauliac pronounced: 'Some people appease [cancer's] treachery and wolfish fury with a piece of scarlet cloth, or with hen's flesh. And for that reason, the people say that it is called "wolf", because it eats a chicken every day, and if it did not get it it would eat the person'.[11] Unlike the 'famous Cancer doctor' described by Turner, most early modern medical practitioners believed cancer to be wolfish in an analogical rather than literal sense. However, the association was a powerful one, which continued from the medieval period well into the eighteenth century. Turner himself, despite scoffing at the notion of cancer as *literally* a wolf, freely admitted the resemblance between this creature and the disease 'for that it is, say some, of a ravenous Nature, and like that fierce Creature, not satisfy'd but with Flesh'.[12] The perceived connection between the devouring behaviour of the wolf and the progress of malignant cancers was so engaging that 'wolf' was used as a synonym for cancerous disease from as early as the thirteenth century.[13] Indeed, the term became so established that some seventeenth-century authors even complained that it was being used too indiscriminately, when it ought to specify a cancer on the legs.[14] Often, but by no means exclusively, practitioners did employ this criteria, using 'wolf' to mean cancer of the legs and thighs. Why this should have been the case remains unclear. It may have been a reflection of the hunting patterns of the wolf, leaping for the back legs of its prey. It may also have been a simple case of utility to find another word for these leg cancers, since the disease was so strongly associated with women that the word 'cancer' often held an unspoken suffix 'of the breast'.

The use of the cancer-wolf analogy in early modern discussions of cancer was widespread and sustained. Perhaps unsurprisingly, then, this vivid image had far-reaching roots. In non-medical writing, and particularly in religious and moralistic texts, the wolf was often connected with anxieties about human frailty and integrity. Such fears are most visible in the rhetorical uses of that animal in the Bible, a source familiar to virtually every early modern English citizen. Genesis 49:27, for example, threatened that 'Benjamin shall ravin as a wolf: in the morning he shall devour the prey, and at night he shall divide the spoil', while Jeremiah 5:6 and John 10:12 depicted the animal in similarly fearsome terms. Throughout such representations, the image of the wolf as a ravenous beast preying upon the faithful flock was foremost: Ezekiel 22:27, for instance, compared the princes of the corrupt house of Israel to 'wolves ravening the prey'. As well as savage power, the wolf was associated with deceit and false appearances. Matthew

7:15 advised the faithful to 'Beware of false prophets, which come to you in sheep's clothing, but inwardly they are ravening wolves'. In the often febrile religious climate of the early modern period, biblical images of the wolf as a fearsome and deceitful predator remained powerfully relevant for many writers of religious or moral polemic. In his 2010 *Animal Characters*, Bruce Thomas Boehrer identifies the wolf as a popular symbol of deception in early modern culture, augmented by the presence of three wolf fables in William Caxton's influential 1483 edition of Aesop.[15] Furthermore, the continued presence of wolves in many Catholic countries after their extinction in Britain in the fifteenth century, and the omnipresent threat of their return to native shores, made this creature a ready metaphor for the perceived Popish threat.[16] In the seventeenth century, Edwards notes that '[t]he figurative wolf in Milton's works consistently represents those with Romish allegiances or inclinations, promoters of superstition, arch-hypo-crites, and rapacious predators'.[17] Milton, she argues, seemingly aligned those church-destroyers with Romish churchmen who lived luxuriously whilst members of their congregation starved.[18]

Inevitably, the interchange between religious and medical rhetoric cut both ways, and numerous writers of polemic soon began recycling the wolf image in ways that explicitly drew on its status as a 'cancer animal'. In the late sixteenth century, for instance, the popular preacher Henry 'Silver-tongue' Smith drew upon moralistic and medical writings when he informed his congregation that '[covetousness is]...like the disease which we call the Wolfe, that is always eating, and yet keeps the bodie leane'.[19] Such writings tended to dwell in particular on the insatiable hunger which was deemed to characterize both actual and bodily 'wolves'. For example, a moralistic poem by seventeenth-century poet Charles Cotton directly echoed Smith when characterising ambi-tion as 'the minds Wolf, a strange Disease, / That ev'n Society [satiety] can't appease' ('Contentment' 1.51–2).[20] By evoking the image of the self 'eaten up' by uncontrollable urges of greed, jealousy or pride, these texts played to an anxiety also identified by Erica Fudge in relation to lycanthropia (werewolves). Writing about lycanthropia, argues Fudge, often dwelt on the humanity or otherwise of the werewolf, debating the disturbing possibility that the creature, being without conscience, was temporarily inhuman (tellingly, inhumanity also extended to atheists, and sometimes to Catholics).[21] Tales of the eating cancer-wolf likewise conjured an image of the wolf undermining, then taking over, the body, diminishing the victim's moral or physical substance. From spiritual, psychological and physical perspectives, therefore, wolves were consist-ently associated with the extinction of the self.

The uses of the 'cancer-wolf' in both medical and literary early modern texts thus show clearly that this image was one shaped by multiple discourses. For medical practitioners, the wolf was an appropriate metaphor for malignant disease and a widely used piece of cancer terminology. On very rare occasions, it was even a 'real' bodily interloper. Poets, playwrights, moralists and clergymen, meanwhile, found in the cancer-wolf an image well established enough to be bent to diverse purposes, underpinned by biblical rhetoric and vivified by contemporary medical doctrine. For all groups, the wolf and cancer were images which readily coincided to describe deception and threat, since both wolves and malignant tumours were characterised by their ability to remain hidden while wreaking destruction. Furthermore, both the wolves described in preachers' sermons and those delineated in medical textbooks threatened to undermine one's humanity, whether spiritual or physical. While the cancer-wolf image never achieved the scientific credibility or cultural saturation of the cancer-worm, its repeated and varied use across genres demonstrates the degree to which early modern people apprehended cancer as a vicious, ravenous and unpredictable threat.

3.2   The worm

3.2.1   Cancer-worms, science and medicine

If the wolf represented the devouring force of cancer, the worm – by which I mean the variety of caterpillars, centipedes, maggots and worms that seem to function in the same way in early modern medical texts – stood for a more insidious kind of malignancy.[22] The image worked in a broadly similar way, with worms imagined as literally involved with cancer and employed as analogies for the disease. However, the worm proved a more popular zoomorphic image, and one with quite different connotations.

The cancer-worm differs most from the cancer-wolf in the extent of linguistic entwinement between disease and creature. Where the term 'wolf' was adopted by medical practitioners because the animal that word describes behaved similarly to a devouring cancer, the cancer-worm concept similarly originated from perceived creatural similitude, but then evolved into a term – 'canker-worm' – which came to designate both cancer-causing parasites and horticultural pests.[23] At one level, the logic behind this evolution is clear. Bodily and horticultural canker-worms clearly shared a *modus operandi*: namely, consuming their 'host' while remaining hidden from view. Harris has briefly described this connection in 'The Canker of England's Commonwealth', where he argues that notions of cancer having 'ontological agency... doubt-less contributed to the emergence in the fifteenth century of the term "canker worm" or simply "canker", to designate a parasitic caterpillar'.[24] In the following century, he contends,

> Through a process of reverse influence, 'canker' the parasite arguably began to affect popular perceptions of 'canker' the disease... Instead of implying an internal humoral disorder, the now multivalent 'canker' more readily suggested a hostile, even foreign organism.[25]

Harris's analysis focuses on the use of 'canker' in economic and dramatic, rather than in medical, texts and contends that during the early modern period, cancers became perceived as 'distinct, hostile organisms, extra-neous to the body rather than produced by it'.[26] His model of recip-rocal influence between horticultural and medical terms, facilitated by rhetorical uses of 'canker', is undoubtedly astute. Nonetheless, that model may flatten the full complexity of this exchange by underplaying medical sources. As evidenced in this chapter, the perceived biological peculiarities of worms in the early modern period allowed for a model of cancer-worm that might be 'distinct' from the body without being an external agent in the way Harris describes. Indeed, medical practitioners never identified the cancer-worm as entering the body from outside, and belief in the inter-personal spread of cancers was, as Chapter 4 discusses, highly atypical in this period. In other words, it was not simply the case that the linguistic development of a horticultural 'canker-worm' in the fifteenth century single-handedly effected the conceptual development of cancer-worms. As I shall demonstrate, biblical, cultural and scientific discourses all had a significant, and hitherto unexplored, role to play.

In order to examine the cancer-worm concept in more detail, one may begin, as with the wolf, at the 'extreme' position of imagining this creature to have literally taken up residence in the body. In this case, however, and for reasons which shall become clear, this position did not represent the end of a spectrum of beliefs, but rather occupied a central location. Many medical practitioners from across the early modern period firmly believed that they had witnessed worms living in, and being extracted from, cancerous ulcers. In 1687, for example, medical practitioner William Salmon reported that

> [a] certain Emperick did cure many Cancers by this one medicine: He took Worms, called in Latin centum pedes, in English Sowes; they are such as lye under old Timber, or between the Bark and the Tree. These

he stamped and strained with the Ale, and gave the patient to drink thereof morning and evening. This medicine caused a certain Black Bug or Worm to come forth, which had many legs, and was quick, and after that the Cancer did heal very quickly with convenient Medicines.[27]

Unlike the story of the wolf discovering its head from within an ulcer, Salmon's anecdote went into detail about the emerging creature and its normal habitat.[28] He took pains that every reader should understand that his description corresponded to what they had seen for themselves under rocks and in damp logs. That specificity brings to life the emergence of cancer from the dank, dark places of the body, offering the reader a vivid image of the disease's progress which was, as discussed later, in line with both biblical and contemporary scientific discourses, and thus adding to the credibility of the account. Interestingly, this passage was an almost verbatim repetition of a tale from D. Border's *Polypharmakos Kai Chymistes*, published in 1651.[29] The 36-year gap between the two testifies both to the power of this image and to the way in which knowledge circulated between texts apparently distant from one another, though the origin of the anecdote remains obscure.

Salmon's story was unusual in offering such a gruesomely detailed image of a creature emerging from a cancerous ulcer, but the premise of his tale was a credible one, which materially influenced therapy for cancers. In printed medical texts and manuscript receipt books, cancer remedies repeatedly promised to 'slea the worme', with one writer suggesting that an application of herbs and butter could tempt worms from a cancerous sore, so that one might 'plucke [the dressing] awaye sodainlye and it will drawe wormes out of it'.[30] Other practitioners, both lay and professional, employed crushed and powdered invertebrates of various kinds in their cancer remedies, clearly seeking to effect a cure by sympathy, or 'like against like'.[31] Moreover, unlike tales of the wolf emerging from the body, belief in cancer as *literally* a worm (or worms) was not necessarily considered unscientific, but seems in some cases to have been absorbed into theories of cancer as espoused by the period's most eminent practitioners. In 1714, Turner, who had related (and discounted) the extraordinary story of the wolf 'peeping out' from within a cancerous ulcer, vigorously asserted the existence of cancer-worms as 'too notorious to want Proof', especially since tiny creatures living in the body could now be observed with the microscope.[32] He added that '[t]he famous *De Mayern* takes Notice also, that he observ'd in the cancerous breast cut from a Woman, some Thousands of Worms'.[33]

This, he argued, explained why 'perhaps the Progress of the Corrosion is sometimes stopt, by applying the Flesh of a Chick, to which these Animals stick, leaving the coarse for the finer Food'.[34]

Turner appealed to new and old medical scholarship in this passage. Belief in the profusion of tiny 'living Creatures' in the body was undoubtedly augmented by the use of that relatively new and exciting technology, the microscope, which allowed one to perceive a world of organisms invisible to the naked eye.[35] Meanwhile, the time-worn popularity of the 'meat cure', as described earlier, seemed to provide practical affirmation of the existence of eating creatures in cancers. As Turner relayed, the cancer-worm theory was thus 'notorious' among 'Learned Men'. Even the most comprehensive works on cancer, such as Dionis's *A Course of Chirurgical Operations*, gave credence to the cancer-worm theory, noting that

> [s]ome believe, that the ulcerated Cancer is nothing else but a prodigious Multitude of small Worms, which by little and little devour all the flesh of the part: What made room for this Opinion, is, that with the Microscope we have sometimes discerned some of these Insects in Cancers; and that putting a bit of Veal on the Ulcer, the Patient has felt less Pain; because, say they, these Worms then feeding on the Veal, leave the Patient at rest for some time.[36]

Such descriptions of a 'multitude' of worms in the flesh highlight the possible origins of the cancer/worm connection. Many early modern citizens would have witnessed at first hand the consumption of carcasses or rotting meat by maggots, and the descriptions here seem to align the cancer patient with these objects. It is also entirely possible that cancer patients with extensive and poorly treated ulcers did find their wounds to become infected with fly larvae, so that worms could be seen at the site of the disease, microscopically or with the naked eye. Indeed, MacInnes contends that during the early modern period, worms in humans, intestinally and in wounds, were 'not pathological, or even unusual, but an expected occurrence'.[37]

Furthermore, contemporary experiments in biology affirmed the potential of worms to appear in the most unexpected of places. MacInnes and Matthew Cobb have separately demonstrated that well into the eighteenth century, it was widely believed that worms could be spontaneously generated by organic matter including plants, mud, manure, hair, wood, flesh and even dew.[38] Accordingly, lurid reports circulated of such creatures appearing, post-mortem, in the body's innermost chambers.

In 1658, for example, a vernacular translation of *The Theater of Insects*, by Thomas Moffett, was appended to Edward Topsell's popular book of zoological observations, *The History of Four-Footed Beasts and Serpents*.[39] Containing some medical material, but clearly intended to entertain and educate a mixed readership, it devoted 17 pages exclusively to the consideration of worms in living human and animal bodies, asserting confidently that worms could breed in numerous spaces of the body, including the heart, and moreover that they might be spontaneously generated from the humours.[40] Still more sensationally, a seventeenth-century text entitled *Vermiculars Destroyed, with an Historical Account of Worms* provided numerous examples of worms found in all parts of the human body, some of extraordinary size or with features such as forked tails.[41] The author also provided readers with instructions for seven experiments via which they could see for themselves the extraordinary ability of worms to be generated from meat, dead snakes, leaves, wood, dust and skin.[42] Such texts indicate that, as in the medical community, public interest in worms was piqued by the popularisation of micros-copy in the mid-seventeenth century.[43] However, as I shall argue, they may also be viewed as part of a wider and much older fascination with body-worms in medicinal contexts.

Contrary to Harris's assertion that cancer-worms necessarily appeared as external agents entering the body from without, both imaginative and medical literature thus suggests that early modern readers appreci-ated some varieties of body-worms as, in MacInnes's terms, 'something latent within the very thing being consumed ... in a real sense, part of the individual'.[44] In large part, this notion was built on empirical foundations and in particular on the rise of microscopy. Underpinning and working alongside these observations, however, was another set of assumptions. Bodily worms generally, and cancer-worms in particular, were creations of a rich cultural and religious history which positioned that creature as a cause, a symptom, and a punisher of weakness and sin.

3.2.2   Worms and corruption in religion and culture

In the Bible, worms – perhaps more than any other creature – appear poised to undermine humans' fragile dominion over nature and misplaced self-importance. Canker-worms may strike at any time to destroy crops and bring about famine.[45] King or pauper, when one dies, 'the worm is spread under thee, and the worms cover thee' (Isaiah 14:11). Moreover, the worm may take on an active role as the punisher (and occasionally the cause) of humanity's sins. According to the scriptures, the undying worm of conscience endlessly tortures the souls of those who have angered God. It has also provided generations of clergymen with a vivid punitive image to impress on their congregations.

From as early as the fourteenth century, it is clear that religious writers seeking to represent the moral tortures of the worm of conscience viewed that creature as analogous to worms which lived in, and gradually devoured, the physical body. Writing on Chaucer's 'Physician's Tale', Harley finds the worm to have been 'frequently invoked in the fourteenth and fifteenth centuries... consistently regarded as an agent of severest torture'.[46] Medieval churchmen warned that 'the "curselyngs... shuln be cast doun into helle... Venemous wormes and naddris [adders] shul gnawe alle here membris withouten seessyng, and the worm of conscience... shal gnawe the soule"'.[47] Like a cancer, these worms devoured one from the inside, and the trope persisted for hundreds of years as poets and polemicists embraced the idea of being literally 'eaten up' by guilt.[48] Just like the pain inflicted by cancers, these tortures were inescapable precisely because they originated inside oneself. Notably, descriptions of the conscience worm gnawing and biting sinners also conflated eating parasites with sharp-toothed vipers. This association between worms and snakes was common in the early modern period, when authors frequently used the terms 'worm' and 'snake' interchangeably, or described worms as 'viperous', venomous or serpentlike.[49] Moreover, the connection between worms and snakes inevitably had implications for how the cancer-worm would be perceived. On the most basic level, snakes had visible fangs, and associating snakes and worms thus lent extra bite (quite literally) to descriptions of the latter creature. Furthermore, Gordon Williams has shown that the worm, which he describes as 'synonymous with *Snake*', was commonly used as a byword for the penis in early modern literatures.[50] Given that cancer was sometimes characterised as a monstrous pregnancy, was deemed 'venomous' and was believed by some medical practitioners to result from venereal infection, it seems clear that the 'semantic freight' of both worms and serpents was brought to bear upon conceptualisations of cancerous disease.[51]

Why were the cancer-worm and conscience-worm images so abiding and widespread, capturing the imagination of so many different audiences? It is clear that these images' correlation with real experiences of intestinal parasites had a part to play, as did the prominence of worms and snakes in the Bible. In addition, I believe it is worth considering just how enduring the human fascination with bodily worms might be. In an article on the supposed presence of worms, newts, snakes and frogs in the body, Gillian Bennet argues that such creatures have, for over 400 years, provided a 'language for sickness'.[52] Indeed, she contends, that language continues to the present day, as evidenced by the Western public's fascination with human parasites.[53] However, even Bennet understates the antiquity of this strange allure. If one looks to discussions of pre-Christian languages and societies, it is evident that fascination with worms in the body, and as a source of sickness, was not exclusive to Judaeo-Christian cultures. Thomas R. Forbes's investigation of early medieval folk medicine, for example, cites charms which are possibly adapted from pre-Christian forms and seek to drive the worm from the body.[54] Looking even further into the past, Watkins's *How to Kill a Dragon* discusses at length both the place of the dragon-slaying myth and its use within a medical context across Proto-Indo-European (PIE) language cultures. With the dragon, as Watkins explains, linguistically and imaginatively transformed into the serpent or worm, 'slaying the worm' in medical terms became a 'mythographic basic formula' across a number of PIE languages – all of which, of course, far predate the early modern period.[55] This formula, frequently expressed through healing charms or poetics, tended to focus upon the 'expulsion' of the worm creature.[56] Furthermore, the formula was linked to another which translates as 'overcoming death', such that, as Benjamin W. Fortson summarises, 'the words used as a vehicle for the serpent-slaying myth ... [encapsulate] not only that myth, but a whole complex of cultural notions pertaining to the slaying of (or by) a monstrous opponent, the struggle of order against chaos, and rebirth'.[57] More work remains to be done on the translation of pre-Christian motifs of illness into Christian contexts, but it appears that, even unconsciously, those early modern writers who employed the worm image accessed an ancient tradition of healing poetics and anxiety about bodily worms.

Conclusion

Zoomorphic characterisations of cancer provided early modern writers with a memorable and flexible mode for imagining a disease which seemed to devour the body in which it was situated. The most extreme iteration of cancer's 'creatural' qualities was, as we have seen, the belief that this disease literally consisted of a worm or wolf present in the body. Interestingly, it appears that this view of cancers as 'parasitic' did not preclude an understanding of the disease as humoral in origin. Even those writers who indicated that they believed cancer might literally consist of creatures inhabiting the body also wrote of the role of melancholy and *atra bilis* in causing cancerous tumours. This ability to subscribe to two seemingly opposed theories of pathology may be viewed as a facet of the broader intellectual flexibility which allowed early modern medical practitioners, as my Introduction suggests, to assimilate aspects of Paracelsianism into medical models that remained broadly humoral. Further along the spectrum, both medical and non-medical writers seized upon these creatures' devouring activities as an apt analogy for the terrifying experience of degenerative disease, drawing as they did so upon the cultural freight that had surrounded images of the worm and wolf for hundreds, even thousands of years.

The impulse to characterise cancer as a creature attacking the body has never gone away, though that 'creature' may now be imagined in less specific terms. James Patterson identifies cancers in the nine-teenth-century imagination as 'uninvited beasts which surreptitiously ganged up on the body', while to this day, fundraising drives, books, research articles and charities continue to exhort audiences to 'kill the beast'.[58] Given the abiding popularity of this rhetoric in the face of (or perhaps in response to) modern medical understandings of cancer which emphasise minute cellular changes, it is hardly surprising that early modern people, confronted with a deteriorating patient and a growing tumour, concluded that the latter was quite literally eating the former. As explored in the coming chapters, this conclusion materi-ally influenced how medical practitioners treated people with cancer and shaped dramatic, politic and poetic renderings of that disease. Through zoomorphism, cancer would be viewed as more hostile than other equally mortal diseases, an evil to be expelled from the body at almost any cost. What makes the worm and wolf images particularly interesting, however, is that they are not simply distillations of the 'devouring' and 'enemy' tropes. Rather, the biblical, imaginative and scientific freight attached to those creatures allowed them to combine – albeit sometimes uneasily – the image of an external creature attacking the body with the sense that the attacked person was in some form responsible for the generation and sustenance of that 'creature'. It was this tension between internal and external which made worm and wolf images such a rich vein of poetic inspiration, and which we shall continue to see at work throughout this book.

4

Cancerous Growth and Malignancy

malignant, adj. and n.

1. a. Disposed to rebel against God or against constituted authority; disaffected, malcontent. *Obs.* 1542—1659 (...)

2. Evil in nature and effects; baleful, harmful, gravely injurious. Formerly also of material substances, plants, etc.... poisonous, deleterious (*obs.*). 1564–1977

3. a. Originally (of a disease): potentially fatal; extremely severe; exceptionally contagious or infectious; incurable. Now chiefly (of a neoplasm): having the property of uncontrolled growth (...) 1568–1993...

4. a. Characterized by malignity or intense ill will; keenly desirous of the suffering or misfortune of others. 1592–1988.[1]

Early modern writers on cancer variously framed the disease as a humoral imbalance, a monstrous progeny or an invading worm. On one thing, however, they were universally agreed. Cancer was characterised, even defined, by malignancy. Moreover, as this definition from the *Oxford English Dictionary* (*OED*) indicates, 'malignancy' was in this period a term with religious, social and political significance, of which the biological phenomenon of uncontrolled growth was only one part. In this chapter, I shall examine how cancer was constructed as malignant in medical, political and cultural discourses. Early modern medical practitioners were, I argue, keenly aware of cancer's malignancy in what we might call a clinical sense; that is, the ability of cancerous tumours to grow and metastasise. To explain this disturbing ability, some writers tried to understand cancer using existing models of poisoning and contagion, attempting to rid the disease of its mystery. In early modern parlance, however, cancer's ability to spread was commonly viewed as a facet of its malignant nature, not the sum thereof. In the interchange between medical and politic or polemic texts, malignancy was constructed in more diffuse terms, as the cruel and evil driving force which impelled cancers to overspread both natural and politic 'bodies'.

At present, little scholarship exists on the meanings of 'malignancy' in the early modern period. Unlike certain other terms such as 'contagion' or 'poison', which have been recognized as having both somatic and figural resonance, 'malignancy' is most commonly treated by scholars of polemic or dramatic literature as denoting a generalised sort of evil, with little attention paid to its medical usage. In addition, while several authors have explored sixteenth- and seventeenth-century medical theories of infection and contagion, none has yet written at length on how early modern people conceptualised the spread of illness within the body – 'malignancy' in a modern sense. Despite these restrictions, scholarship on the *inter*personal transmission of illness in this period does provide a useful model for considering the *intra*personal spread of cancer. Among many others, Kevin P. Siena, Vivian Nutton and Rebecca Totaro have noted how medical anxieties about the infectious potential of bodily fluids, breath, touch or even sight operated in relation to seemingly non-medical discourses about gender roles, national morality and travel.[2] In each case, models of infection slipped easily between medical and non-medical discourses, 'draw[ing] even the moral and emotional phenomena to which they were applied back into the circle of medical analysis'.[3] Medical terminology and theory was not only turned to rhetorical purposes in non-medical texts, but was in turn shaped by these imaginative reworkings.

Understanding the way in which medical and imaginative or polemic texts shaped one another relies in large part on recognising the correlation between natural and 'politic' bodies in early modern writing. From both literary and historical perspectives, it has been shown that large communities such as the church or the state were frequently imagined as composite bodies, dependent on complex relationships between 'organs' of production and regulation. Naturally attendant on such an image was the possibility of imagining dysfunction in the body politic in corporeal terms. Sarah Covington, Colin Milburn and David Harley, among others, have pointed out the rhetorical utility of describing a nation as wounded, syphilitic or requiring physic.[4] Recent scholarship on the designation of monstrous births as symptomatic of socio-political ills, or the politically motivated reimagining of skin complaints, underscores the degree to which the analogy cut both ways, with politics mediating bodily experience.[5]

This chapter discusses both senses of 'malignancy': that of neoplasmic growth and the broader sense of 'ill will'. In the first section, I discuss how medical practitioners and patients attempted to understand the ability of cancer to spread through the body by relating it to other phenomena including poisoning and contagion. In the second section, I consider how cancer's growth was understood as indicative of the disease's 'malignancy' in a broad sense: its evil, rebellious quality. Positioned in this way, 'malignant' cancers became an apposite image for talking and thinking about any person or group felt 'likely to rebel against God or authority', with that dissenting spirit feeding back into discourses of the disease's pathology.

4.1   Cancerous growth

In the twenty-first century, 'malignancy' is most often used to describe the propensity of cancer to grow and spread throughout the body. Early modern medical practitioners, as I will show, used 'malignancy' in a broader sense. Nonetheless, they too were keenly aware that cancer was an invasive disease. Why, they asked, did some cancers grow so large that they developed into ulcers, while others disseminated to diverse parts of the body?

Over several hundred years, medical practitioners of all kinds devoted much attention to describing cancer's disturbing tendency to increase and spread. A 1651 edition of the popular *Directory for Midwives*, for example, delineated the progress of breast cancer as 'a little tubercle, no bigger than a pease, [which] ... grows up by degrees, and spreads out roots with Veins about it', while *The Compleat Midwife's Practice* stated that cancers 'sometimes remain for two years together, no bigger than a Bean; afterwards it grows to be as big as a Nut, then to the bigness of an Egg; and after that increasing to a larger size'.[6] Such descriptions followed a widespread trend when they compared the incipient tumour with familiar objects distinguished by their potential to grow or bring forth life; elsewhere, medical practitioners described tumours as growing from the size of a pea, nut or bean, to that of a Crown, egg or even a small melon.[7] The primary object of interest in such discussions was the single cancerous tumour which grew larger and larger. Less commonly, however, medical writers noted that tumours might also appear in relatively distant parts of the body – in modern terms, metastasise. For instance, the anonymous writer of *An Account of the Causes of Some Particular Rebellious Distempers* added the following note of caution to their promises of a cure for incipient cancers:

> If a Cancer in the Breast proceeds from malignant Humours or corrosive Salts in the Blood, it is generally incurable ... or if in some it

should seem to yield, or indeed seem to be cur'd, while it proceeds from those corrosive Humours, they many times breed again, and break forth afresh, either in the same place, or in some other part of the Body.[8]

As this account demonstrates, medical practitioners frequently viewed tumours which arose in diverse places as separate maladies caused by the same corrupt humour, rather than a single disease which had migrated within the body.[9] Nonetheless, they recognized that cancers which recurred once were likely to keep doing so. In the case of the man with a tumour the size of a melon, the attending surgeon recorded that after he had treated the patient, he was informed that he had been treated before for a tumour in the same location, in that case as big as a cherry.[10] This knowledge, he wrote, 'gave me Reason to apprehend a Return of the Distemper, tho' it never happened'.[11]

Medical practitioners emphasised cancer's ability to grow and spread more than almost any other facet of its pathology. This was largely for practical reasons. It was obvious that the body could not sustain a tumour which grew exponentially, and tumours which rapidly expanded were thus understood as posing the greatest risk of a morbid and painful cancerous ulcer. This development was much feared by medical practitioners, and presumably their patients, with good reason. Cancerous ulcers were almost impossible to cure, and were known as painful, stinking and disgusting, provoking lengthy and largely identical descriptions throughout the early modern period. In 1597, for example, Peter Lowe asserted:

> [The ulcerated cancer] is an ulcer round horrible, having the lippes thick, harde, inequall, sordide, turned over, cavernous, evill favoured, of colour livide and obscure accompanied with many veines full of Melancholick blood, voyding a matter virulent, sanious worse than the venim of beastes, subtill waterie, black or red.[12]

Almost identical accounts of 'pestilent', 'loathsome' and foul-smelling ulcers can be seen in the 1698 *The Compleat Midwife's Practice* and Pierre Dionis's 1710 *A Course of Chirurgical Operations*.[13]

Certain features of the cancerous ulcer remained important throughout such discussions. The darkened veins which designated a growth as cancerous in its first diagnosis reappeared here as a means of making clear this malady's difference from other kinds of ulcer. The 'Lips' of the wound, with all their disgusting characteristics, brought to mind both ingestion and excretion, framing the ulcer as at once a discrete organism and a grotesque parody of natural function.[14] Ulceration could happen with relatively small tumours, particularly if they were poorly treated. However, they were most strongly associated with tumours which grew rapidly, giving the impression of breaking through the skin from within. In therapeutic terms, there was almost universal consensus on the mortality of ulcerated cancers. Numerous practitioners pronounced that in such cases, 'nothing but Death is to be expected' and palliative care was the recommended course.[15] So significant was the ulceration of cancers that many medical practitioners treated ulcerated (or 'exulcerate') and non-ulcerated cancers separately within their texts, setting out different prognoses, treatments and other advice for the two complaints from the outset.[16]

Medical practitioners thus saw an accurate and timely assessment of cancer's growth and spread as essential to predicting the outcome of the disease, after which they might either decline to treat it, treat it with palliative methods only or amend their therapies according to the aggressiveness of the complaint. It was, for example, deemed very important not to use emollient or suppurating medicines on a tumour that grew rapidly and might ulcerate, while surgery was judged an appropriate course for discrete lumps but not for those suspected to extend deep into the body. In some cases, it was seen as a victory simply to keep the cancer from spreading too rapidly. Reporting the illness of 'Mrs. Ladd' to her uncle Henry More in 1674, 'Dr. Clark' announced that though it remained painful, the lady's breast tumour was not discernibly larger, 'which makes me hope that the Medicine is proper for it'.[17] In addition to these practical considerations, discussions of cancer's growth were imaginatively important. Growth, and the ulceration associated with it, were *the* factors by which cancerous tumours could be distinguished from more benign lumps and bumps, and although cancerous growth and malignancy were not the same thing, the former was understood as a vital component of the latter. Accounts in which the expanding tumour appeared to possess an exponential capacity for growth implied the 'taking over' of the body by a cancer that was ontologically separate, such that at some crucial tipping point, the victim's human substance, and with it their life, would be eclipsed by the mass of the tumour. That distinctly spatial emphasis is found repeated in, for example, Wiseman's description of cancer's propensity to 'spread and invade the neighbouring parts', or Jane Sharp's note that malignant tumours 'daily increaseth with roots spreading', both of which used metaphors (militaristic and arboreal) to scale up the space occupied by the disease mass.[18]

Medical authors universally agreed that a propensity to increase was definitive of cancerous disease. Exactly *how* cancers grew, however, was another matter entirely. The majority of medical practitioners seemingly paid little attention to this question, attributing cancer's capacity for growth to its 'malignancy' in a broad sense, as discussed later. In several cases, however, writers on cancer sought a different sort of solution to this problem, with recourse to models of illness which were more established and of which medical practitioners felt they had a better understanding. For want of a better term, I shall call these models 'aetiologies', though in early modern parlance, the immediate causes of disease could hardly be separated from their pathologies or 'natures', which were in turn far removed from anything we might recognize as such today.

Most prominent among early modern aetiologies of cancerous growth was what we may broadly term the 'poison model'. This model proposed that cancers emitted some venomous or poisonous substance which caused either neighbouring or distant parts of the body to become sick in their turn.[19] *The Compleat Midwife's Practice*, for instance, asserted that '[t]he cancer is a venomous tumour', and several works by eminent practitioners throughout the early modern period seemed – at least, at certain points – to draw a similar conclusion, describing the matter believed to emanate from cancers as a 'corrosive and malignant venome'.[20] Supposedly poisonous cancerous liquids could be emitted from a tumour or an ulcer and were strongly associated with foulness, bad smells and pain. In 1597, for example, Jacques Guillemeau described such secretions as 'thyn corrupt matter, more vile then the poison of any wilde beast, most abhominable both for abundance and smell, and the payne is continually pricking'.[21] Over a century later, describing the effects of advanced cancer upon a female patient, Browne poignantly recorded the way in which

> the Ulcer became more corrosive, and spread its Venome all over her Breast, even to her Arm-pit; and after this, the whole Arm on that side being therewith inflated, she became dispirited with the great pains she daily felt, and lived some short time in this miserable condition, till Death put a stop both to her pain and her days.[22]

Though the poison model of cancerous growth could not make the disease any less cruel, it effectively united emotive characterisations of cancer's 'vile' effects with a familiar clinical aetiology. Unsurprisingly, then, this model appealed to writers seeking a satisfying explanation for cancer's growth within the framework of humoralism. It may also have been augmented, from the mid-seventeenth century, by the claims of contemporary scientists that some venoms were produced by the rage or fear of the venomous creature. In a lengthy text on natural philosophy, Robert Boyle related an experiment in which he had fed various parts of a snake to a passing dog and found that the dog suffered no ill effects. This, he proposed, supported the general observation that a snake's venom *'consists chiefly in the rage and fury wherewith they bite, and not in any part of the Body, which hath at all times a mortal property'*.[23] A schema which viewed poisons as chemical substances, generated by qualitative emotional states, allowed medical practitioners to credit cancer's capacity for growth to poison without abandoning long-held ideas about the disease's being 'evil'. In addition, the poison theory, particularly when expressed in terms of 'venom', fitted closely with imaginatively potent characterisations of the disease as a creature independent of the patient, whether that was a worm, a rabid wolf or a monstrous product of the troublesome womb. In non-medical writings, descriptions of cancer as venomous or poisonous were certainly less prevalent than in medical or scientific treatises. Nonetheless, a few authors adopted this aetiological model, which usefully allowed one to imagine cancer as both a local and a systemic malady. One anonymous invective against duelling, for example, described the practice as a 'wild and inverterate [*sic*] Cancer, that has diffused its Venom thro' all the liquid Mass'.[24] In this image, the ferocity and resistance to cure of the 'wild' Cancer was combined with the ability of poison to permeate the whole body.

For medical and polemical writers, the poison model thus appealed as a mode of thinking about the perplexing spread or growth of disease in the natural or politic body. Imagining a cancerous poison or venom, however, also raised its own problems. This theory implied that the tendency toward aggressive growth characteristic of cancers inhered in a material substance, and some medical practitioners even believed that this substance could be isolated by scientific experiments. In 1670, for instance, one anonymous writer asserted that the 'Malignity and Poison' of cancer 'discolours the purest Metals, if touch'd with it'.[25] In an altered version of essentially the same idea, William Beckett proposed in 1711 that cancer was caused by disturbed lymphatic juices, such that 'if we express a Juice from some of the *Cancerous Mass*, and hold some of it in a Spoon over a Fire, there immediately flys off a small Vapour, and the Remainder hardens not unlike the White of an Egg boil'd'.[26] On one hand, it was implied that, if the 'venom' of cancer could be isolated in this way, then it could be understood and treated. As Miranda Wilson notes, poison was popularly believed to be a predictable method of death, and this was amplified in contemporary drama, where poisoners were depicted as being able to choose the day and even hour of their victim's demise.[27] Attributing cancer's growth to poison thus promised a similar degree of control over this disease. On the other hand, experiments such as the above also seemed to show that the poison responsible for cancerous growth could exist outside of the body and could thus be transferred from one body to another. Though an uncommon perspective, this disturbing possibility was raised by an extraordinary story related in *An Account of the Causes of Some Particular Rebellious Distempers*, which is worth repeating at length:

> Those inveterate and dangerous Cancers but seldom happen, and is frequently more from want of timely and proper Applications than the Nature of them; for they are oftentimes aggravated and enraged, and the Humour, by wrong Applications inwardly and outwardly, made corrosive and sharp, as we frequently find it to be; and the Humour is [...] as subtle, quick and penetrating as Poison it self, as will appear from the following Relation, which a Surgeon tells us happened upon himself...Mr. Samuel Smith, one of the Surgeons of St Thomas's Hospital in Southwark, who at the cutting off of a large Cancerated Breast, had (after the Breast was off) a Curiosity to taste the Juice, or Matter contain'd in one of the little Cystis's or Glands of the same, which he did by touching it with one of his Fingers, and then tasting it from the same with his Tongue, the Taste of which he protested did immediately like a Gass, pierce through the whole substance of his Tongue, and passed down his Throat not less sharp or biting than Oyl of Vitriol, Spirit of Nitre, or Aquae Fortis, or some vehement Catheretick, or Caustick Salt, and altho' he presently spit out, and wash'd his Mouth with Water, and that oftentimes, and also with Wine, and drank presently very freely of Wine after it, yet could not get rid of the Taste thereof, but it continued with him, and brought him (who was a very strong Man) into a Consumption, or wasting pining Condition, attended with several other ill Symptoms, which in a few Months after killed him, the Taste thereof never going off from his Tongue to his dying Hour; and that the Taste of the Juice, or Matter of that Cancerated Breast, he declared upon his Death-bed, and near the last Moments of his Life, to be the true and only Cause of his languishing Condition and Death.[28]

Questions about power and gender raised by this curious incident are discussed in Chapter 6. Here, however, we can note the unusual way in which the anonymous account identified a malign 'essence' capable of causing consumption in one person and cancer in another. Furthermore, imagining cancer in these terms did not prevent the author from crediting the disease with a degree of sentience. Though apparently identifying an efficient cause for cancer, *An Account* continued to use language which construed the disease as acting with evil intent, able to be 'aggravated and enraged' by attempts at cure. In short, this seemingly new solution to the mystery of cancer's spread through the body raised the same old fears and created some new ones to boot.

The story of Samuel Smith's demise was undeniably compelling. Marjo Kaartinen notes that it was retold in at least five medical treatises spanning more than a century.[29] However, the notion that cancer was transmissible by poisoning generally failed to gain much traction among either medical or non-medical writings on the disease. The reason for this failure seems to have been simply that cases such as Smith's were extremely rare. Some 40 years after *An Account* recorded this event, Beckett's *New Discoveries Relating to the Cure of Cancers* revisited the tale. Beckett revealed that he had done the same thing, having 'diluted some Drops of the Juice in several Spoons-full of fair water, till at Length, not finding any Inconveniences from it, I came to the Juice it self.'[30] Beckett's experiment, however, left him unscathed, and he concluded that the death of Mr. Smith was due not to the corrosiveness of the cancer 'juice' itself, but because its offensive taste and smell disturbed Smith's own 'Animal Juices' and disordered his whole body.[31] Smith's experience simply did not hold true in Beckett's experiments, and neither did it fit with Galenic theories of disease, which focussed on humoral (im)balance. This incompatibility need not necessarily have been an obstacle to the idea's adoption – the case of zoomorphism has shown how medical practitioners could ignore 'violations' of the Galenic model in order to accommodate useful tools for thought – but the fact remained that poison generally offered only a reformulation of the original causative gap between cancer's substance and behaviour. Inadequately supported by contemporary theory to be adopted as a useful mode of explaining cancerous growth, cancer-poison was, for the most part, an image quietly assimilated into broader conceptualisations of the disease as intrinsically foul.

The poison model of cancerous growth and metastasis never became orthodox in early modern medical texts. However, the impulse to match the perplexing disease of cancer with more familiar somatic phenomena can be seen in numerous medical works from throughout the period. Particularly prominent was the idea, not dissimilar to that of cancer-poison, that cancer was pathologically related to infectious diseases, particularly leprosy and venereal pox. Those two diseases were themselves often understood as related to one another. As Marie McAllister has shown, contemporary speculation on the origins of pox sometimes traced the 'foul disease' to sex between a leprous man and a menstruating woman.[32] Elsewhere, sufferers of the two diseases were linked by shared facilities or common therapeutics.[33] Few scholars, however, have noted that leprosy and pox were in turn understood to have characteristics in common with cancer. In 1703, Browne's *The Surgeons Assistant* stated confidently that 'Leprosy also ariseth from the same cause and matter [as cancer]; and they are seen only to differ in respect of the part in which they consist'.[34] The notion that leprosy and cancer differed in degree rather than quality was, according to Demaitre, a widely held notion dating from the eleventh-century writings of Avicenna.[35] Five hundred years later, the link was still going strong, with Philip Barrough's 1583 *The Method of Physick* categorizing cancer as a variety of 'lepry'.[36] Both the supposed humoral imbalance and the skin lesions characteristic of leprosy appeared to align the disease with cancer, such that leprosy could be considered 'cancer of the whole body'.[37]

In a similar manner, descriptions of venereal pox during the early modern period frequently highlighted the similarity between ulcers or sores created by this disease and those associated with cancer. In a text on pox entitled *Little Venus Unmask'd*, the Dutch physician Gideon Harvey described a venereal infection as yielding 'crusty black sanious devouring Ulcers or Soars, [which] did eat holes into the Yard, like Cancers, yea some of those Cancers or Shankers made but three or four Suppers in Devouring the whole Virge [penis]'.[38] Harvey clearly understood 'Cancer' as a separate disease which produced effects 'like' those of pox, but he was happy to appropriate the term, as well as the zoomorphic 'Devouring' associated with cancer, to vivify his description of pox sores on the genitalia. In doing so, he followed an established trend: as we have seen, 'canker' was often employed in early modern parlance to describe undifferentiated ulcers, including of the genitals, and Harry Keil has observed that 'cancre' (or variants thereof) was likewise sometimes used as an indiscriminate term for venereal lesions in medieval surgical texts.[39] Cancer and pox were further united by the use of mercury ointments and 'salivation' as cure for both diseases.[40] In line with the widespread notion that benign tumours or inflammations could become cancerous if they were treated incorrectly, several medical texts also described cases in which the authors suspected that venereal disease had 'caused' the patient's cancer, though they seldom provided a theoretical basis for this suspicion.[41]

Speculation on the relationship of cancer to syphilis and leprosy was clearly motivated by pragmatic observation of their similarities and by, as has been noted of the supposed leprosy/pox connection, an 'urge to translate the mysterious new disease into a familiar one'.[42] For medical practitioners struggling to understand how cancer grew and spread, it also offered new terms in which to imagine that phenomenon. As Browne argued in the early eighteenth century,

> A Cancer...that is exulcerated, may be allowed to have in it a great share of Contagion; it being bred from the same humour as the Leprosy is; and I know nothing that can contradict this my opinion, unless you allow, that a Contagion cannot be referr'd to any single Part, but must be communicated to the whole Body.[43]

Contagion, the force which was understood to spread leprosy and pox from one body to the next, might also be imagined as driving the *intra*personal spread of disease, so that a cancerous tumour 'infected' adjacent parts of the body. Later in the same text, Browne would reiterate this view and insist that since leprosy and cancer were of the same 'temper', and leprosy was catching, one could naturally conclude that cancer was contagious on a smaller scale.[44] The comparison posed some difficulties: as Browne acknowledged, most people believed that contagion could only affect whole bodies, not parts thereof, and he was the only author to explicitly depict malignancy as a variety of contagion. Nonetheless, several sixteenth- and seventeenth-century medical practitioners used the terms 'infection' or 'contagion' in a more casual sense as shorthand for cancer's potential or actual spread. Advising on cancer surgery, for example, Paré stressed to his readers that one should cut away 'whatsoever is corrupt, even to the quicke, that no feare of contagion may remaine, or be left behind'.[45]

Imagining cancer as contagious did not necessarily make it any easier to treat. After all, neither leprosy nor venereal pox was reliably curable, and medical practitioners struggled to understand the different modes of transmission for various infectious diseases.[46] Moreover, there was no suggestion that understanding cancers as intrapersonally contagious could help one to halt their spread within the body. There was also a more disturbing twist to this theory. While the vast majority of practitioners adjusted the explanatory model of contagion to describe the spread of cancers *within* the body, for a few individuals, the reverse was true, and the model began to reshape their perceptions of cancerous disease. The results of this perceptual shift can be viewed in two unusual tales from Beckett's 1711 *New Discoveries*, which are worth exploring at length.

Beckett's first account was passed onto him by an acquaintance, and concerned a tradesman's wife in Nottingham suffering with breast cancer. 'Her Husband', wrote Beckett, 'was of Opinion he cou'd relieve her by sucking it; accordingly he put this Method in Practice, in hopes without doubt he cou'd effect a Cure, by drawing the Cancerous Matter out of the Nipple'.[47] This strategy did not work, and the woman died soon after, but after two months her husband experienced a swelling in his upper jaw. Turning (unsurprisingly) from surgeons who recommended that he have the swelling and part of the jaw bone cut away, this tradesman pursued a course of gargles 'and such inconsiderable remedies', but was eventually obliged to consent to the surgeons' original suggestion; too late, for the cancer then spread over the mouth and nose.[48] Becoming 'so frightful an Object, and the Stench that continually proceeded from the Parts... so offensive', the patient removed himself to a garret, where he died.[49] Similarities to venereal pox in particular are powerfully evident in this account. Suckling at the breast was a recognised means by which nursing infants could contract the disease, such that catching pox from a wet nurse was a danger frequently pointed out by advocates of maternal nursing.[50] More generally, the use of this case to illustrate, as Beckett put it, '*Whether Cancers are Contagious, or not*' seemingly relied on the fact that the tradesman's disease appeared localised to the spot at which he had had contact with the original cancer, rather than diffused through the body as in accounts of poisoning.

The importance of localised 'infection' to the construction of cancer as contagious was even more emphatically stressed in Beckett's second account, of cancer transmitted skin to skin. In this 'very odd Accident', a poor woman with ulcerated breast cancer continued to share a bed with her two children.[51] Shortly afterwards

one of 'em, a Girl about five Years old, began to be afflicted with a small painful Tumor in one of her Breasts, which encreasing to near the Bigness of an Egg, became Livid, and entirely *Cancerous*; the Mother died some time after, and the Child did not survive her; but the other Child continu'd well. Several Surgeons gave their sentiments of this Case; some thought it to be an Hereditary Indisposition, but considering the Mother had no appearance of a *Cancer* before, or at the Birth of the Child... [I believe] it was contracted by Contagion, seeing the Position of the Child's body was such in Bed, that that Part of it which was affected was almost always disposed to rub against the Dressings soaked in Matter; (for I understand the Mother took but very little Care to change them often). Now it is not at all probable, that the malignant *Effluvia*, which continually pass off from the *Cancerous Mass*, and the putrefied Matter, can dispose a Person at any little Distance to be afflicted with the like Disease, for then the other Child wou'd have become a Sufferer; but it may happen in some extraordinary Cases, where the corrupted Fluid has attain'd an exalted Pitch of Malignity, to communicate some of its more active Particles to the Blood and Spirits... but this cannot happen unless the matter be very malignant; and be suffer'd, by the negligence of the Patient, to come to an immediate Contact, with a Part of the Body of the other Person.[52]

As in the story of Samuel Smith's poisoning, an extreme version of the malignancy threat was here represented by cancer's transmission from one body to another, scaling up the spread of tumours from between members of the body to members of society. The danger was exacerbated by moral turpitude – in this case, the 'little Care' of the afflicted mother, which was seemingly more reprehensible than Smith's fatal surgical 'curiosity'. Nonetheless, Beckett's case for contagious cancer was timid at best. Kaartinen writes that '[q]uite a number' of early modern medical practitioners believed cancer to be contagious, but this appears to be truer for the mid- to late-eighteenth century than for the period under examination here.[53] Rather, Beckett's emphasis on the exceptional circumstances which surrounded this contagion by cancer reflected the singularity of his account. In general, belief in cancer as contagious was precluded in this period by a distinct lack of cases such as the above. In the vast majority of writings on cancer during the sixteenth, seventeenth and early eighteenth centuries, contagion was not even mooted as a possible cause, and, as Samuel Smith's tale demonstrates, medical practitioners did not generally approach cancer sufferers as contagious or dangerous; the 'noli-me-tangere' ('do not touch') label applied to some cancers was understood to protect the welfare of the patient, whose tumour could be irritated by manhandling, rather than that of the touching practitioner.

Early modern medical authors and their audiences were fascinated by the ability of cancer to grow and spread through the body. Cancers grew unpredictably, sometimes to astonishing proportions. They reappeared after seemingly having been cured, and, most worryingly, they broke through the skin to create painful, morbid ulcers. Moreover, their ability to 'invade' the body in these ways was troublingly mysterious. Practitioners who described cancers as poisonous or contagious had one aim: to make cancer less frightening by showing how it worked. These attempts provoked discussion about the causes of and possible cures for cancer, but in general they failed to exert much influence on medical practice. Strikingly, however, the inconsistencies and omissions of these models show how attempts to understand exactly *how* cancer moved through the body neither superseded, nor clashed with, literary and medical constructions of cancer as purposefully 'malign'. Instead, they found themselves positioned somewhere between rhetorical and material understandings of the disease.

4.2 The character of malignancy

While a select few medical practitioners speculated about theories of contagion and poison, they were always in the minority. Most of those who encountered cancer, in text or in person, perceived the malady's spread through the body in more general, and arguably more disturbing, terms. Cancerous growth was understood as indivisible from the broader quality of 'malignancy': a property which helped account for the painfulness of cancer and its resistance to cure, as well as its propensity to spread, and which was viewed as intrinsic to the disease in a way quite foreign to modern conceptualisations of illness. In this section, I discuss how for early modern people, the malignancy which underlay cancer's spread through the body was largely indistinguishable from the malignancy of villainous individuals or factions as represented in literary, religious and polemical texts. This concept traversed the permeable boundary between literal and figural representation such that 'malignancy' became a potent and protean idea: a product of somatic experience, medical theory *and* literary imagination.

Even for expert medical practitioners, cancer was a difficult illness to diagnose. As discussed earlier, and in Chapter 1, medical textbooks from across the early modern period emphasised the diminutive size of incipient cancerous tumours, which were described as 'hard to be discovered', growing and damaging the body but impossible to find, let alone treat.[54] Correspondingly, of all the aspects of cancer's pathology, the ability to remain 'secretly hidden' was perhaps that which most fired the non-medical imagination, proving crucial to literary constructions of 'malignancy'.[55] In political and poetic rhetoric, the canker-worm, an image which often mixed characteristics of cancers and horticultural cankers, typically described a hidden threat. Karen Edwards, for example, notes of worms in John Milton's poetry: 'That it destroys slowly and in secret is what turns a caterpillar or insect larva into a canker-worm, rhetorically speaking'.[56] The same is often true of Shakespeare's works, which repeatedly use 'canker' as a byword for weaknesses or vices concealed even 'in sweetest bud'.[57] In drama and verse, therefore, the hiddenness of cancer often stood for ideas within an individual, or individuals within a society, whose harmful influence went undiagnosed.

The implied threat from such 'inward' cancers was not only their concealment *per se*. Rather, it was the way in which secrecy permitted the growth of a sickness which would, upon discovery, threaten the natural or social body. This aspect of cancerous disease was a point of particular interchange between medical and popular texts, as medical accounts presented cancer's 'emergence' from the interior of the body in dramatic terms. In particular, the word 'discovery' was frequently used by medical practitioners to describe the coming to light of a previously unseen cancer, either as a tumour which had grown to become palpable and visible, or, more commonly, a cancerous growth which had broken the skin to create an ulcer.[58] Relating the progress of a breast cancer tumour, for example, Gendron described how 'the growth of them at last pierce the Skin, and discover the Cancerous Mass', later adding that facial cancers might similarly 'discover themselves'.[59] Such descriptions neatly united the contemporary senses of 'discovery' as literally removing the cover from an object and figuratively 'disclosing to knowledge' something previously secret.[60] Moreover, the narrative of a purposely 'secret' disease which was suddenly 'discovered' played to constructions of cancer as a *dramatis persona* with its own, predetermined, agenda.

Using loaded terms such as 'secrecy' and 'discovery', medical discussions of the progress of cancerous disease frequently emphasised what seemed like the independence of this malady from the body in which it was found. Early modern medical practitioners of all kinds repeatedly implied that in some sense, cancer did not simply respond to the conditions of the body, like other illnesses, but rather 'aimed' to reach its apotheosis in the breaking out of a cancerous ulcer and the death of the patient. Whereas in twenty-first-century terms, 'malignant' or uncontrolled growth is understood as a result of the cellular pathology of cancer, for early modern medical writers and their audiences, it made more sense to reverse that equation, and view malignancy as the intrinsic 'character' which determined the pathological effects of

cancerous disease. As such, cancer was frequently and vehemently identified as evil and cruel; as Dionis asserted, 'the most terrible of all the
evils which attack Mankind':

> though Wars and Plagues kill in less time, they don't yet, to me, seem
> so cruel as the Cancer, which as certainly, though more slowly, carries
> those afflicted to the Grave, withal causing such Pains as make them
> every day wish for Death.[61]

Throughout the early modern period, cancer was characterised as
purposefully evil. The anonymous 1670 *An Account*, for example, noted
that a cancerous tumour 'grows big of a sudden, and discovers its evil
Nature by the grievous Symptoms that appear, and as it increases in
bigness, it increases in malignity'.[62] Bonet similarly described cancerous
ulcers as having an 'evil' and 'Malignant' disposition which purposely
'eluded' cure.[63] Again and again, the disease was deemed 'cruel and
horrid', 'cruel and terrible', 'fierce', 'stubborn' and 'indomitable'.[64] These
terms often operated in a multivalent sense. Describing a disease as 'evil',
for example, could indicate that it was deemed likely to have a poor clinical outcome or to cause further complications. However, pathological
effect was in these cases virtually indivisible from ontological cause, so
that cancer was deemed evil, cruel and fierce – in short, malign – in a
way that encompassed moral 'intent' and somatic consequences.

The characterisation of cancers as 'evil' had far-reaching consequences for how that disease was experienced imaginatively and physically. As described in Chapters 5 and 6, both medical practitioners
and their patients bore in mind the supposedly intractable character
of cancers when making decisions about pharmaceutical and surgical
interventions. Furthermore, these notions of cancerous malignancy
surfaced throughout the early modern period in non-medical literature, where they interacted with discussions of villainy, violence and
deception. Non-medical writers often seized upon the idea of a secret
or hidden cancer or canker as an analogy for concealed moral vices or
subversive individuals.[65] Similarly, many authors adopted the notion
of cancers or cankers as initially minor disruptive elements working
toward a destructive apotheosis. Matching their medical counterparts,
these culminations were often violent in character, associated with
damage to the body politic, and on occasion to the individual body
too. Wither's 'Opobalsamum Anglicanum' is an apt example to which
to return here. Casting parliamentary corruption as a 'cancer', the poem
adeptly plays upon the multivalent senses of 'cancer' to warn that this

malady 'will effect the Bodies overthrow: / Or, els (beside much trouble, griefe, and cost) / Occasion many Members to be lost' (l.67–74).[66] When the growing influence of malignancy is not 'interrupted', it is argued, chaos follows, as the poet alludes to the multiplicity of his image. The 'Bodies' – that is, the individual body and the figurative political body – will be overthrown both in the sense of succumbing to illness and that sense (in 1645, never far from the poetic mind) of political revolution or breakdown. Playing still further on the bodily degeneration associated with cancerous disease, the author's warning of 'Members' to be lost clearly puns upon that word as denoting both Members of Parliament and members, or parts, of the body – parts which might, in turn, be lost as a result of violent civil unrest.

As discussed earlier, comparisons between sickly natural and politic bodies were a commonplace of early modern literature. Cancer, however, provided a particularly useful tool for thinking about treachery, treason and moral failure. In Wither's poem, the author's invocation of a mutinous element which was hidden, corrupted the surrounding parts, and was both of and hostile to the 'body' necessitated that it should be *cancer* specifically that 'sickened' Parliament, and lent a visceral, violent tinge to its possible 'overthrow'. The same use of cancer's unique pathological and 'behavioural' characteristics was repeated elsewhere in both persuasive and dramatic literature. Gerrard Malynes's 1601 treatise on the 'canker' of foreign trade, for instance, construed the national 'body' as being overwhelmed by economic disadvantage in the same way that a cancer sufferer was overcome by their growing disease, and ended with 'the politike body of our weale publike ... overtaken', in an image that played on cancer's literal mortality.[67] Likewise, John Fletcher's drama *The Faithful Shepherdess* (1608) described the lecherous 'Sullen Shepherd' character as 'like a Canker to the State', who mimicked the location and action of bodily cancers by 'eating with debate / Through every honest bosome' (5.3).[68] That all these texts imagined cancer's destruction on a national scale was no more a coincidence than the characterisation (discussed in Chapter 5) of the disease's resistance to cure as a 'rebellious' act. Cancer, which seemed malignant in an ontological sense, yet was unmistakably generated by the body, was perceived as not only a cruel disease but a traitorous one, turning against that which nourished it. This aspect of malignancy can be seen used to powerful effect in both religious and civil contexts. In his essay on medical metaphors, for example, Harley notes that '[a]fter 1640, when sects such as the Baptists and Quakers started to proliferate, orthodox Calvinists were quick to assert that "False doctrine is like a Cancer or Gangreene, it frets all that is sound and in the end killeth"'.[69] In a similar manner, clergyman Thomas Adams described those who stole from the Church as lying 'in the bosome of the Church; as that disease in the brest, call'd the Cancer, vulgarly the wolfe: devouring our very flesh, if we will not pacifie and satisfie them with our substance'.[70] The ferocity of the 'wolfe' was important to Adams, but equally significant was the placement of the traitor or cancer in the 'bosome' of the institution, central to the body and associated with nurturing and re-productivity (unlike another 'eating' disease, gangrene, which primarily affected the body's extremities).

In each of these cases, cancer's intimate connection with the body which it destroyed was essential to the translation of malignancy from the individual body to the body politic. In addition, both medical and non-medical texts occasionally drew attention to subtler aspects of the similitude between bodily and social malignancies. In particular, the ability of cancer to spread through the body unchecked, and the unpredictable rate at which it did so, proved valuable to its rhetorical capital as a byword for violent dissent. Texts such as the anonymous treatise against duelling *An Account of the Damnable Prizes in Old Nick's Lottery* placed particular emphasis on the fact that this 'wild and inverterate cancer' of upper-class society outpaced as well as outfoxed attempts at a cure, noting that as it 'laid hold of every nobler part with its deadly Claws' it would only 'spread the more and faster' when met with opposition.[71] Moreover, the author's concern with the speed at which a moral 'cancer' might spread once again aligned with wider concerns about the political and social impact of individual movement across the country. As Andy Wood has pointed out, the seventeenth century saw the first use of 'Mob' as shorthand for describing disturbingly *mob-ile* plebeian crowds. In the politic 'body', controlling the movement of 'malign' people and ideas was felt as a vital, and increasingly difficult, task.[72]

The meaning of malignancy as 'likely to rebel against God or authority' was thus influenced by the somatic experience of cancer's progress, but in turn fed back into how cancerous malignancy was reported and experienced. Moreover, what it meant to 'rebel' depended, rather conspicuously, upon what or who one deemed an authority. While at the turn of the seventeenth century Shakespeare cast 'cankers' as acting against royal authority, by the time of the Civil Wars, 'Malignants' had come into use as a term applied by parliamentarians to Royalists.[73] Whichever way the political wind might blow, the cruelty and morbidity of cancerous disease ensured that 'malignancy' remained a useful image with which to discuss power, duplicity and destruction. Furthermore, by looking at medical and non-medical texts in tandem, it becomes evident that the latter also influenced the former. The conceptualisation of malignancy may profitably be viewed as a circuit upon which the somatic experience of cancer and the social disorder related by texts using the malignancy image were two opposite points. Each relation of civil or religious disobedience as cankered or 'malignant' fed back into medical discourses to furnish those writers with the language in which to describe the bewildering and frightening experience of encountering malignant cancer. In turn, increasingly vivid accounts of somatic experience recirculated to set up cancerous malignancy as a powerful and apt metaphor for the description of troubling or violent disorder in the body politic.

Conclusion

For early modern people, 'malignancy' was a term rich with somatic and social associations, describing more than the clinical fact of neoplasmic growth with which the word is associated today. A large part of what was denoted by malignancy in medical texts was the terrifying ability of cancers to spread through the body or recur after their apparent cure. In trying to understand these phenomena, some medical practitioners tried to model cancerous growth using theories which were, by the standards of the day, biomechanistic in approach. These attempts loosely prefigure the move which would take place during the eighteenth and nineteenth centuries toward attempting to understand cancer according to new iatrochemical and germ theories.[74]

Visible throughout even the most radical medical theories about cancerous growth, however, was the abiding sense that cancers spread and took over the body simply because this was central to their nature. 'Malignancy', as it described the disease's spread and its resistance to cure, was absolutely intrinsic to the disease. The diagnostic criteria which marked out a cancer from a benign tumour, such as heat, pain and discolouration, were likewise deemed signs of cancer's malignant nature. Moreover, 'malignancy' was also understood as the force which brought those grievous symptoms about, such that it seemed that cancers *were* malignancy in action – its bodily manifestation. It was this sense which facilitated the association of cancerous malignancy as a mode – still present in twenty-first-century discourses – of talking about moral ills, or those which spread through the politic or religious body. Rebellious subjects could easily be imagined as, like cancerous tumours, the physical embodiment of an intangible urge toward destruction and disruption, characterised by a troubling illimitability and unpredict-ability. This vision of malignancy was a multi-authored creation, in which the social and political concerns of the age were attached to the somatic experience of, and medical anxiety around, a disease which unfailingly provoked horror, apprehension and curiosity. 'Malignancy', therefore, was neither a medical term borrowed by literature, nor a metaphor adopted by medical practitioners, but a term of true intertextuality.

5

Wolves' Tongues and Mercury: Pharmaceutical Cures for Cancer

Early modern patients diagnosed with cancer were positioned at the centre of debates about gender, the nature of disease, anatomy and the humours. More practically, they also found themselves with a malady that was often painful and disfiguring and had the potential to end their lives. Confronted with such an illness, what was to be done? The following two chapters examine how cancer sufferers, and the medical practitioners who attended to them, attempted to stem or reverse the effects of this disease. I will argue that, in their most potent forms, cancer treatments continued the conceptual separation of patient from disease which was visible in zoomorphic and anthropomorphic descriptions of cancer's character. In so doing, they diminished the patient's role in their own cure, while foregrounding an adversarial relationship between medical practitioners and 'rebellious' cancerous tumours. Throughout the early modern period, cancer treatments provoked fierce debate over both the nature of disease and the proper limits of medical intervention.

In this chapter, I investigate non-surgical therapeutics for cancer. These are loosely defined as those remedies which did not involve the dreaded 'Knife or fire' in cutting or otherwise penetrating the flesh (though as I shall show, this did not mean that such remedies could not cause fissures, intentional and otherwise, in the patient's skin).[1] Such 'cures' were incredibly varied, ranging from strict diets, to unguents such as oil of frogs (made by baking the creatures with butter in their mouths), to powerful purges of hellebore or senna, and toxic caustics including arsenic and mercury.[2] They were also employed by diverse parties: although surviving sources primarily document those cures prescribed by professional medical practitioners, many empirics, apothecaries and lay people had their own opinions on how best to cure a cancer. Despite the apparently disparate nature of these materials, a thread can be traced through cancer therapeutics. Prescriptions can roughly be graded, as I have divided them here, into orders of severity, from the merely unpleasant, to the acutely dangerous, with the most radical therapies accompanied by elaborate rhetoric and impassioned debate. Departing from treatments based on regimen and rebalancing the humours, which involved the active participation of the patient, increasingly complex and potent pharmaceutical interventions focussed less and less on the individual with cancer, and foregrounded the zoomorphised or anthro-pomorphised tumour. At length, therefore, both medical practitioners and patients faced a decision: in order to kill a cancer, how far were they willing to go?

The most comprehensive look at early modern cancer treatments is currently provided by Marjo Kaartinen's *Breast Cancer in the Eighteenth Century*, and the medical landscape that text describes is in many respects similar to the one I explore in this chapter.[3] Kaartinen notes, for example, the continuity of lay and 'professional' therapies for cancer and the incomplete distinction of curative from palliative remedies.[4] Moreover, it is clear that many of the therapies employed in the eighteenth century were ones which remained unchanged over several hundred years, even dating back to the medieval period. Receipts made from ingredients such as plants, animal dung and grease were passed down in domestic receipt books and through printed texts of various kinds from the sixteenth into the eighteenth century.[5] Likewise, lead and mercury waxed and waned in popularity as cancer cures, but remained in use for well over three hundred years.[6] In other aspects, however, it is clear that the later eighteenth century in particular was characterised by a preponderance of exotic cures for cancer with no equivalent in earlier texts, including the ingestion of lizards, use of electrical therapy, and application of carbonic acid to tumours.[7] These innovations evidently depended on alterations in medical theory, and in socio-economic circumstances, which were unique to the later reaches of the early modern period. Aside from Kaartinen's study, early modern treatments for cancerous disease have seldom been investigated at any length. In her *Female Patients in Early Modern Britain*, Wendy D. Churchill briefly describes the way in which women with breast complaints often delayed seeking medical attention until pain or debility made it absolutely necessary, fearing the painful methods of 'cure' offered by both physicians and surgeons.[8] Luke Demaitre's essay on cancer in the medieval period also notes the reluc-tance of medical practitioners to interfere with cancers, as well as the particular use of prescriptions from the *Dreckapothecke* – that is, various

kinds of excrement.[9] In his lengthy study of the history of cancer therapeutics, Siddhartha Mukherjee evocatively characterises early modern
cures for cancer as 'an intricate series of bleeding and purging rituals to
squeeze the humours out of the body as if it were an overfilled, heavy
sponge'.[10] However, these works have paid relatively little attention to
the way in which remedies for cancer reflected beliefs about the nature
of the disease, or medical practitioners' relationship to the malady.

This chapter also looks to scholarship on other, more studied, diseases
in order to contextualise some of the methods and ingredients employed
in the treatment of cancer. In particular, works on venereal pox have
provided valuable details about the unpleasant side-effects of mercury
'cures' which may help to explain why this course was such a controversial one in the treatment of cancer.[11] As noted in the Introduction to
this book, recent scholarship has also foregrounded the protean nature
of the early modern medical marketplace. Iatrochemical methods
were, as I discuss, incorporated into medical systems which remained
broadly humoralist in both theory and praxis.[12] Likewise, the differenti-
ation between varieties of medical practitioners – physicians, surgeons,
apothecaries and itinerant medicine-sellers among others – was often
problematic. Particularly outside London, the line between authorised
and 'empiric' practice, as well as between areas of specialisation, was
blurred. Patients might pick and choose from a broad range of practitioners depending on their budget, complaint, locale and personal
preference.[13]

The methods by which medical practitioners attempted to treat cancer
were diverse, complex and, in many cases, incompletely recorded. With
many medical textbooks listing multiple cures, it is likely that more
than one avenue of therapy was pursued at any one time, so that, for
example, a patient might undergo purging, apply daily salves or lotions,
and maintain a modified diet, more or less simultaneously. Broadly
speaking, however, it is clear that many early modern patients and
medical practitioners subscribed to the intuitive approach of beginning
treatment with mild therapies, and moving on to increasingly violent
ones if the disease failed to respond. This is the schema upon which the
chapter is divided, and along which I trace a corresponding conceptual
shift leading to the exclusion of the patient, as an individual and an
agent in their own recovery, from a drama played out between medical
practitioners and cancer.

The first section of the chapter therefore discusses recommendations
for the regimen of the cancer sufferer – their diet, the administering
of medicinal purges to expel excess humours and bloodletting. Such

prescriptions, I argue, were based on an understanding of the disease as humoral in origin and emphasise the responsibility of the patient for their own physical well-being. The second section looks to internal medicaments, unguents and salves which were specific to cancer. Such therapies increasingly treated cancer sufferers as interchangeable, becoming more invested in combatting cancer as an ontologically independent complaint. In the third section, I look at those medicines – usually applied to ulcers and tumours on the body's surface – with corrosive properties, primarily arsenic and mercury. Unpleasant, and frequently dangerous, such therapies were highly controversial, but, tantalisingly, seemed to promise an 'eating' force to rival that of the malignant tumour.

5.1   You are what you (don't) eat? Combatting cancers with diet and regimen

On an unrecorded date in the mid-seventeenth century, the physician John Fernelius wrote to his colleague Simon Pietre for advice regarding a tricky case of 'cancerous wenns' in the armpit of a young woman.[14] Pietre's reply tells us much about the way in which medical practitioners approached this disease, and is worth citing at length. The letter began with a brief recipe for ointment to be applied to the wens (sub-dermal lumps), with a warning against using any strong medicines. The bulk of the letter, however, was taken up with detailed instructions for the woman's regimen, which, Pietre argued, ought to include regulation of diet, medicinal purges and bloodletting:

> the whole body of this ingenuous damosel ... is tender and dry, as I understand by her Father, it must be gently handled. And therefore it must be purged with Cassia Fistula, Diacatholicon, or King Sapors syrup newly made, half the saffron being left out. Which let her take twice or thrice in a month, with whey wherein Epithymum and fumitory have been infused. And because her nature seems inclined to breed melancholick juice, even of the best meats, through fault of her Livers distemper; we must fight against that juice with a syrup made of juice of bugloss, Borage, Caume, Endive, sweet prunes, whereof let her take amornings with boyled water. To the same intent Asses milk will be good, which let her use every morning with a little sugar. At the approach of spring and Fal, her body being purged, let her left basilica or median veine be opened, and take two smal porringers of blood. Finally, make an issue in her left Arme, neare the muscle Deltois. This summer time let her frequently use a bathe of sweet

fresh water, to correct the driness of her body. Moreover, let an opiate be made for her of Conserve of Violets, Lillies, Roses, Bugloss, Borrage, Citron peel, Confectio alkermes, that by the use thereof the malignant force of the melancholy juice may be amended and the patients natural strength restored.

Let all her diet and course of life tend to moistness, and moderately to cool; refusing all meats that breed melancholick juice.

Let her use ptisan [tisane] instead of wine, or a decoction of coriander with Raisons. And when the heat of the weather shall be more remiss, you shall order her wine well allaid with water, which in this extremity of summers heat I do not allow.[15]

Pietre's recommendations found favour with Fernelius, who saw fit to include them in his 1662 *Select Medicinal Counsels*. Moreover, they reflected precisely the belief of many contemporary practitioners: that cancer was a disease with humoral origins, which might be cured by redressing bodily imbalance.

Widely believed to have its origins in humoral dysfunction – namely, the burning or stagnation of black bile – cancer represented an ideal candidate for redress by adjustment of the 'non-naturals', diet and bodily regimen. In turn, regulation of non-naturals was perhaps the most widespread form of medical prescription during the early modern period, and, as Andrew Wear has observed, an idea firmly embedded in the nation's cultural consciousness.[16] Insightful work has lately been written on the importance of regimen in neo-Galenic therapeutics. Jan Purnis, for example, argues in 'The Stomach and Early Modern Emotion' that attention to diet reflected a 'profoundly embodied partnership' between body and mind in early modern somatic experience.[17] Margaret Healy similarly observes that '[i]t is probably true to say that the maxim, "We are what we eat", was never so significant in England until this period', with adherence to certain dietary rules deemed essential for the spiritual and physical health of both individual and country.[18] In his 2002 *Eating Right in the Renaissance*, Ken Albala approaches the relationship of food and medicine from the opposite direction, arguing for a sincere, if sometimes confused, interest in the medicinal effects of food from culinary writers across Europe.[19] While scholars have investigated the early modern relationship between food and medicine from different perspectives – indeed, a comprehensive work on food *as* medicine in this period has yet to be written – they are in agreement on two points; both of which, I will contend, are highly visible in texts dealing with cancer. They concur that

drinks and foodstuffs were thought of as having heating, cooling, moistening or drying properties, by which they could create or redress humoral
imbalance in the consumer. Moreover, they argue, this direct connection
between eating and *being* temperate or intemperate cast food as a mode of
self-determination with intertwined moral and physical consequences.

We have seen in Chapter 1 that cancer was most often conceived of as
a disease of *atra bilis*, a noxious derivative of the melancholy humour.
In their advice on the most appropriate diet and regimen to counteract
or prevent cancers, medical practitioners varied little across 150 years,
returning repeatedly to the recommendations of moderation and avoid-
ance of 'strong' meats found in Galen's *Methodus Medendi* as a means
to quell excess melancholy.[20] In 1583, for example, Philip Barrough
advised that 'among other thinges this is chiefly and principally to be
observed, (namely) that such nourishment be given to the diseased, as
have vertue to refrigerate and moysten, and which doe engender good
and slender juyce'.[21] He went on to specify 'fishes of gravelly places',
egg yolks and poultry (excepting that which 'live in fenny groundes')
as particularly desirable foodstuffs, his descriptions demonstrating the
remarkable specificity with which gamey meats or sea fish were distinguished from their lighter counterparts.[22] The foods Barrough prescribed
were believed to be cool and moist in quality; meals thought not to tax
the digestive system, and, perhaps, ways to tempt a sickly appetite. In
this the physician conformed to Galen's advice that melancholy individuals should 'use meats that are light of digestion'.[23]

This injunction was heeded over and over by medical practitioners
from the sixteenth to the eighteenth century. A 1698 edition of *The
Compleat Midwife's Practice*, for instance, advised a diet of 'cooling and
moistening spoon-meats' for any woman with inflamed breasts that
might turn cancerous, while other writers prescribed spare and bland
food and drink.[24] Any reader confused as to the components of a 'spare'
or 'cooling' diet could turn to those, like Ambroise Paré, who decreed in
detail which foods a cancer sufferer might safely eat, and which should
be avoided:

> thicke and muddy wines, vinegar, browne bread, cold hearbes, old
> cheese, old and salted flesh, Beefe, Venison, goate, hare, garlicke,
> onions and mustard, and lastly all acride, acide and other salt ... which
> may by any meanes incrassate [thicken] the blood, and inflame the
> hum[ours] ... be eschewed. A cooling & humecting diet must be
> prescribed; fasting eschewed, as also watchings, immoderate labours,
> sorrow, cares, and mournings; let him use ptisans, and in his brothes

boile Mallowes, Spinach, Lettuce, Sorrell, Purslaine, Succory, Hops, Violets, Borradge, and the foure cold seeds. But let him feede on Mutton, Veale, Kid, Capon, Pullet, young Hares, Partridges, Fishes of stony rivers, reare Egges; and use white wine, but moderately for his drinke.[25]

Paré's injunction against red meats and strong savoury flavours in favour of white or 'young' meat, green vegetables and fish was typical among his contemporaries. The caution given here against 'thicke and muddy wines' was also commonplace, with Alexander Read later asserting that 'there is nothing more pernicious [for melancholy complexions] than the immoderate use of potent and strong wines, such are all kinds of Sacks, and greeke wines, which exceedingly burne the humors in the masse of the bloud'.[26] Such prescriptions followed the logic of humoral theory to the letter. Moist meats and broths, for example, were believed to counteract the dry melancholy humour which led to the stagnation of blood, and the separation of noxious properties within the blood which 'resembleth the dregges of wine'.[27] Meanwhile, warnings against the evils of excessively strong liquor drew on the advice in Galen's *Methodus Medendi* that a person of choleric complexion should 'fly from Wine and strong Beer as fast as he would fly from a Dragon'.[28] The emphasis placed on avoiding strong alcoholic drinks in texts on cancer thus demonstrates the degree to which choler was felt to be implicated in the transformation of melancholy into *atra bilis*, which in many texts appeared as a process of burning or 'adustion'.

Most dietary recommendations had their roots in Galenic theory and were justified in those terms. However, they also incorporated a degree of moral proscription, resting as they did upon patients' everyday choices around food, drink and physical activity. In her article on 'Sciences of Appetite' in the eighteenth century, Elizabeth A. Williams argues that 'seventeenth-century medical advice was marked by eating anxieties and by medical antagonism toward gastronomic indulgence'.[29] The red meat, strong cheeses and potent wines described as causing cancers fell into that category of 'indulgences', and only the wealthy could afford to eat such items regularly. In particular, medical writers repeatedly identified foreign, especially Greek, wines as dangerous to health, recommending instead watered-down wine or small beer.[30] Indirectly, they thus linked cancer to epicurean or intemperate appetites, seemingly remaining oblivious to the fact that those who could afford their services were by definition likely to be among those few who enjoyed a rich, varied diet. In addition – as so often in discourses about the disease – women were once again marked out as particularly vulnerable. As described in Chapter 2, large breasts, as associated with obesity, were viewed as a risk factor for breast cancer. Women were understood as likely to have more body fat than men because of their more sedentary lifestyles, their lack of self-mastery, which led them to over-eat, and their cold humours, which were inadequate to fully concoct, or 'burn off', rich meals.[31]

As Healy states in 'Bodily Regimen', 'staying healthy had enormous spiritual and moral implications' in this period, in which 'disease had become a culpable and blameworthy affair closely associated with over-indulgence'.[32] Where lifestyle prescriptions for staying or curing cancer extended their reach beyond diet, this moral dimension became even more pronounced. In accordance with Galen's recommendations for choleric complexions, writings on cancer repeatedly warned against strenuous exercise.[33] Exercise could seldom be considered immoral in itself, though it might be less than genteel. However, many medical practitioners extended that proscription to include mental and emotional 'labours', which one had a duty to try and moderate. In 1650, for instance, Read echoed the advice of many of his contemporaries when he advised that 'watching [brooding], immoderate labour and griefe' should be shunned by cancer patients, since, like certain foods, they heated the body and facilitated the creation of *atra bilis*.[34] As discussed in Chapter 2, when viewed alongside the belief that blows or bruises could bring on cancers, such speculation on the dangers of grief and anger takes on a darker perspective, in which spousal violence is tacitly indicated as one way of generating this disease. Not only were women most likely to be the victims of such violence, it was also believed that they had difficulty in controlling their emotions, making them, once again, more vulnerable to the ill effects of melancholy.

Readings of cancer which positioned diet and regimen as crucial to one's chances of surviving the disease might be read as disempowering. One might naturally have a melancholic or choleric disposition which was particularly susceptible to heating by unsuitable foodstuffs. Equally, a hostile home environment, bereavement or other outside factors might bring on the harmful 'watching' and grief which exacerbated the disease – not to mention the fact that 'immoderate labour' was not a matter of choice for many early modern patients. Yet while cures which emphasised the need to balance the humours highlighted certain circumstantial or physical predispositions to cancer, they also stressed the connection between moral, psychological and physical health, and offered opportunities for holistic self-determination – namely, the chance to heal oneself. So firmly was this belief engrained in the mind of medical practitioners that some writers recorded great frustration with patients who neglected their prescribed regimen. Writing in 1711, William Beckett complained: 'I cabbit [cannot] say whether I had more trouble with the *Cancer*, or in endeavouring to oblige my Patient to a strict Observance of some of the non-Naturals she so often err'd in'.[35] Repeating the age-old grievance of medical practitioners, he bemoaned patients 'nor taking so much Care of themselves, as they expect that the *Surgeon* should take of 'em'.[36]

Despite their occasional obstinacy, by careful regulation of diet, and procedures such as purging and phlebotomy, cancer patients could, it was believed, evacuate corrupt matter from the body, redress their faulty humours and help themselves to become healthy again. As such, concoctions designed to purge the whole body of excess humours were a staple of almost every printed medical text and appear as a natural progression from the regulation of the body through food and exercise. A 1662 translation of Lazarus Riverius's medical observations, for example, emphasised the importance of purging before any other avenues of cure were to be pursued, and even proposed that purges could completely cure an incipient cancer:

> where speaking of a Cancer, [Galen] has these words. *I have often Cured this Disease when it was but beginning, but when it is grown large, it cannot be cured without manual operation*; and a little after: *this disease I have (as was said) Cured at the beginning, especially when the melancholy humor was not very thick; for then it easily gives way to purging Medicaments, by which the Cure is effected*; and it is easie to conceive, that these purging Medicaments must purge black choler…Herefore I conceived I must fly to the use of strong Remedies, the chief of which is the Root of black Hellebore, which is most effectual to purge Melancholly.[37]

Black hellebore was a favourite purgative for Riverius; a 1655 translation of the author's *The Practice of Physick* again asserted that 'by giving the Extract thereof twice or thrice, we have somtimes cured a Cancer in the beginning'.[38] Medical practitioners throughout the early modern period placed similar store by the effectiveness of this poisonous substance, often combining it with gentler ingredients such as senna, rhubarb and endives in a broth or tisane.[39] While purges might be less toxic than the concoctions of arsenic and mercury favoured by some physicians, however, they could hardly be considered an easy option. Senna was well known as a laxative, and hellebore was a powerful emetic, potentially lethal in the wrong hands. In this light, Riverius's repeated emphasis on purging the body takes on extra emotional importance. The cancer-causing *atra bilis* was 'conceived' as something more than mere chemistry, appearing instead as a malign progeny to be driven from the body, and the discomfort of purging was recast as a personal 'labour' by which such mal-productions might be expelled.[40]

In the related process of bloodletting, expulsion of ill humours from the body was similarly positioned as an exercise to heal while it hurt. Removing harmful *atra bilis*, and often standing in for interrupted menstrual or haemorrhoidal bleeding, bleeding appeared as a positively intuitive response to illness for a variety of medical practitioners and their patients. Riverius, for example, suggested bleeding 'in the Arm, Anckle, and Hemorrhoid Veins' as an effective means to stay, if not to cure, cancers of the womb, while John Browne believed that judicious bleeding could stay cancers in any part of the body.[41] Such recommendations demonstrate the considerable faith placed in this therapy. As Gail Kern Paster records, many medical writers, well into the eighteenth century, conceived of the circulatory system as moving blood only slowly around the body.[42] In redress to the blockages and stagnation thought to result from this state of affairs, the black blood which medical practitioners often claimed they could see collecting around tumours was supposedly removed by phlebotomy. Bleeding from the haemorrhoids or feet was believed to draw blood away from the cancerous areas of the upper body, starving the tumour of *atra bilis*. Opening the ankle (saphaena) vein in particular was also believed to redress the humours by provoking menstruation (and to procure miscarriage: an action linked pragmatically and figuratively to expelling the mis-conception of a cancerous tumour).[43] As in the case of digestive purges, however, this treatment was not without its dangers. Browne's text contained a stark warning about letting blood around the area of a tumour. 'I have more than once observed in my Practice', he asserted, 'that letting the Patient Blood in the same Arm...on that side the Cancer is fixt, that new Cancers have readily been bred thereupon, and which have many times been more malign, and much worse than the former'.[44] If phlebotomy had the power to move good blood into the area of a tumour, it also had the potential to move corrupting blood into other parts of the body, prompting what we now call metastasis. Medical practitioners also realised the risk posed to a patient's already failing strength posed by bleeding, exhorting readers to 'be cautious' and only let 'as much as the patient can suffer'.[45]

While it presumably deterred some patients, the 'suffering' involved in being bled may be viewed as integral to the perceived efficiency of the procedure as therapy for cancer. In contrast to the unbidden menstrual bleeding that signalled 'woman's inability to control the workings of her own body', Paster argues that 'the control of blood and bleeding exemplified by the phlebotomist's art becomes a key determinant of agency and empowerment' among both sexes.[46] Indeed, much of the bloodletting carried out during this period was self-prescribed as a prophylactic, with wealthy individuals summoning the phlebotomist or barber-surgeon at certain times of year, or whenever they felt themselves 'plethoric'.[47] Even if bloodletting was not the patient's own suggestion, it was a procedure with clear, explicable logic for those familiar with the basic principles of humoralism.[48] While cancer was often frustratingly mysterious in its causes and progress, bleeding offered patients the chance to control their bodily substance in a way that was tangible and visible.

Prescriptions for controlled diets, calm and quiet activities, medicinal purges and bloodletting were among the most common recommendations to appear in medical texts discussing cancer during the early modern period. As we have seen, they held considerable appeal, apparently undiminished by their potential to cause discomfort or even physical harm. Largely self-directed, cures based around regimen offered therapies that were readily understandable to patients, with medical practitioners possessing specialist knowledge – the best places from which to bleed, for example – but no basic insights into the procedures which were not virtually common knowledge for a population steeped in Galenism. Moreover, these cures were, to some extent, tailored to each patient's constitution. As Eve Keller describes, purging, phlebotomy and dietetics were all embedded in a discourse at once holistic and individualistic: holistic, because it foregrounded the interaction of self with environment; and individualistic, because it emphasised the uniqueness of each patient's constitution.[49] Nonetheless, there were downsides to such therapies. Diets, purges and bleeding were 'catch-all' cures, designed to redress humoral imbalance and thus heal the whole body *including* the tumour, rather than to target the cancer specifically. For patients battling malignant tumours, searching for a definitive cure, this was often not enough. They sought more radical means, and in the thriving medical marketplace of early modern England, they found many sources willing to supply them.

5.2   Plantain and wolves' tongues: herbal and animal remedies

Pietre's letter to Fernelius advocated, as we have seen, the regulation of diet and lifestyle above all else. The medical practitioner's first concern, he argued, should be to redress the unbalanced humours which afflicted the whole body, effectively to starve the cancer of the *atra bilis* upon which it was founded. Having achieved this, however, Pietre also recommended applying a more specific cure. 'I conceive you ought to deal very gently with [the tumours]', he wrote:

> nor must you use strong softners or digesters, least they grow worse, but gentle ones, such as is an ointment made of a little diacalciteos dissolved in juice of Plantane and Nightshade, al beaten together in a laden [leaden] morter.[50]

As cures went, this ointment was among the simplest, consisting of three basic ingredients. 'Diacalciteos' most likely refers to chalcitis, an oxide of iron commonly used in medicines of the period. Plantain and nightshade were common plants with respectively soothing and poisonous properties. Finally, the leaden mortar imparted some of its toxicity to the finished mixture. Despite its ingredients, however, Pietre's scant description implied that this was a 'gentle' remedy, perhaps temporarily alleviating pain in the affected area through the mortifying effects of nightshade. Most telling is Pietre's caution to his colleague: 'nor must you use strong...digesters, least [the tumours] grow worse'. In this statement is contained the weight of a belief held by dozens of practitioners treating cancer, that aggressive therapies for the disease caused them to *grow worse* as if in an act of rebellion.

Pietre's fear can be traced back at least into the sixteenth century in English medical texts, and remained current well into the eighteenth. Barrough's 1583 *The Method of Physick,* for example, exhorted the reader to 'make choice of those medicines, which are of a meane force, and of a gentle qualitie'.[51] His recommendation was explicitly tied to a conceptualisation of cancer which imagined the disease in anthropomorphic terms; Barrough believed that 'the malignitie of the evill through...vehement medicines is stirred, and provoked, and made more fierce and savage'.[52] Similarly, in 1651, Nicholas Culpeper's popular *Directory for Midwives* noted that cancer 'hath a peculiar malignity, which is fermented and mad[e] worse with Emollients and suppuraters'.[53] Imagined as semi-sentient, the capricious, ill-tempered cancer demanded to be only 'softly medled with'.[54] How exactly medical practitioners and patients believed that these medicines did 'meddle with' the disease is often unclear. Ingredients for such 'gentle' prescriptions were widely varied, frequently including plantain, rose oil or water, borage, honey, lead, alum, henbane and nightshade.[55] Many medical writers, and in particular the writers of household receipt books, recorded these components, and the method to make their medicine, with no other comment attached than the ubiquitous 'est probatum' ('it is proven'). What is evident, however, is that there was no single cancer-curing herb included in these remedies. Rather, combinations of ingredients were chosen to combat the cancer through a mixture of symptomatic relief and redress of the *atra bilis* which caused the disease. Plantain, for example, was held by Culpeper's *English Physitian* to be a plant of such general usefulness that 'there [is] hardly a Martiall Disease but it cures', and was deemed particularly good for quelling fluxes and easing pain and inflammation, all features of cancerous disease.[56] Roses were likewise credited with a myriad of healing properties, including reducing inflammation, purging choler and strengthening the vital organs.[57] The seemingly counterintuitive inclusion of toxic plants such as henbane and nightshade into cancer remedies was believed, when applied correctly, to assuage pain and swelling.[58] Balancing so many different properties, such remedies could be incredibly complex to prepare, with one cure from Elizabeth's Godfrey's 1686 receipt book listing 42 separate ingredients. This lengthy process, however, was deemed worthwhile when it seemed to produce results: recording the receipt, Godfrey noted that '[this] is the best was ever found out ... cour'd Mrs Finches maide'.[59]

Although in general it was the combination of ingredients which made these remedies specific to cancer, there was one notable exception. Animal products, including various kinds of fat and dung, were common in a range of medicaments for various diseases and were accordingly used in ointments and unguents for tumours. In remedies for cancer, however, some creatures – crabs, certain arthropods (mainly woodlice and centipedes) and worms – were found with a far greater frequency than elsewhere. Pechey, Barbette and Paré were among the many prominent early modern medical practitioners who included powdered crab in their remedies for cancer.[60] Furthermore, they drew upon a long therapeutic tradition. Michael B. Shimkin identifies the ingredient as used similarly during the 'dark ages', while A. Kaprozilos and N. Pavlidis list crab as a main ingredient in plasters and ointments for cancer in ancient Greek texts.[61] The inclusion of crab in cancer remedies was not explained or justified in the texts, leaving us to speculate as to its supposed utility. Given the close association of the crab with cancer, however, it seems likely that crab-based remedies were believed to work on the principle of 'like against like'.[62] This principle is more obvious in relation to the inclusion of less common 'like' ingredients in cancer cures. The German physician Oswald Gabelkover, for example, advised in the late sixteenth century that '[f]or the gnawing Wolfe, or Canker' one should '[t]ake a Wolves tunge, drye it, and beate it to poudre', before making it into a plaster with honey, and then 'wash the disease with wine & strewe of the poudre of the Wolves-tunge therein till such time it be cured'.[63] The difficulty presumably involved in procuring a wolf's tongue testifies to the power it was believed to possess against cancer, also known as 'the wolf'. Belief in the efficacy of 'like against like' is even more visible in this 1651 account, in which a medical practitioner used a variety of 'worm' in medicine to expel the worm of cancer:

> A Certain Emperick did help many cancers, in divers people (that were troubled with them) after this manner. He took certain worms, called in latine Centumpedes, in english sows: they are such as lie under old timber, or between the bark and the trees. These he stamped, and strained with ale, and gave the patient to drink thereof morning and evening. This medicine caused many times a certain black bug, or worm to come forth which had many legs, and was quick, and after that the cancer would heal quickly with any convenient medicine.[64]

Once again the powerful 'like' ingredient required no additions, no combination with other substances to work its cure. The sole purpose of the ale seems to have been as a medium in which the 'Centumpedes' might slip down more easily.

Medicines containing parts of worms or wolves highlight the slippage that occurred in early modern medical discourses between viewing those creatures as apt analogies for cancer and imagining them to be physically involved in the disease. Being less common ingredients than crab, they tended to be discussed at greater length, illuminating more clearly the principles behind these 'like cures like' remedies. First, the 'stamping', crushing or burning of the animal material could be seen as an act which transferred negative feelings about the tumour onto its substitute in the medicine. That is, the annihilation of the crab, wolf or worm ingredient might symbolically stand in for, as well as physically effecting, the anni-hilation of the zoomorphised tumour. Secondly, the spatial emphasis in Border's account implied a different kind of substitution. The worm or 'centumpede' was taken from its dwelling place between the bark and trunk of a tree; a place which, appropriately, recollected the sub-dermal or sub-cutaneous positioning of many tumours. Being bent to the purposes of the empiric through stamping and straining with ale, the reformed 'centumpede' appeared to drive out the many-legged 'bug' from the body, as if only one could occupy that space at any moment.

The harmful cancer-worm was replaced with a similar creature which was beneficial to the patient and, crucially, controllable by the medical practitioner.

Attempts at curing cancer with crabs and worms illustrate the degree to which many pharmaceutical cures tended to treat the cancer as a discrete entity, rather than redressing the humoral balance of the whole body – despite the fact that the authors of these cures did not identify themselves as interested in ontological disease models in an academic sense. This was partly a product of cancer's imaginative construction as ontologically independent of the cancer patient. Such cures were also products of the expanding medical marketplace. As Harold Cook has pointed out, demand for new goods in this period meant that practitioners could make more money selling cures for specific diseases than they could balancing the complexions of a few wealthy patients.[65] In this increasingly competitive environment, a gulf emerged between writers – often university-educated, licensed physicians – who emphasised the difficulty of curing any established cancer, and other medical practitioners, sometimes advertising in newspapers or pamphlets, who promised a quick, cheap and painless cure. Though licensed medical practitioners undoubtedly have the loudest voice in surviving historical documents, there nonetheless remain tantalising glimpses of the prestige achieved by some so-called empirics. In 1714, for example, Daniel Turner described one 'famous Cancer doctor' as a 'villainous empiric', indicating that one might specialise in this disease as other unlicensed practitioners did in bone-setting or cutting for the stone.[66] He advised those who had cancer that they should on no account

> [l]ist[en] after a promised Cure by cosening Quacks, or Cancer-curing Pretenders, who, to my Knowledge, have hasten'd great Numbers of People miserably to their Graves, who might otherwise (and that very tolerably) have spun out a much longer Thread and have kept under this really (so far as I know of Surgery) incurable Distemper.[67]

For their part, the 'Cancer-curing Pretenders' attracted 'great Numbers' of people to their services by promising what Turner felt he could not. Advertisements for internal medicaments or gentle ointments to cure a cancer quickly and painlessly were frequently accompanied by testimonials from satisfied customers enjoying newfound 'Health and Ease'.[68]

Why did these advertisers – some, licensed physicians, but many, apothecaries or 'unauthorised' practitioners – give a prognosis so much more optimistic than that found in medical textbooks? There was certainly profiteering at work and the fact that such sources are self-selecting. Nobody advertises that they cannot cure a disease. Nevertheless, the fact that these drinks or salves were frequently touted as 'universal', curing everything from dropsy to gout, is also instructive. Customers who purchased one of these cure-alls probably did so of their own volition or on recommendation from friends and family, since medical practitioners were understandably reluctant to send business to their competitors. Therefore, they were less likely to have received a formal diagnosis of cancer, such as an examining physician might provide. Their disease may have been less advanced, and they may have been less concerned with whether it was a 'true' cancer (as opposed to a cyst, scirrhus or mastitis) than whether the cure-all managed to relieve it. This also seems to be the case for writers of household receipt books, who had little to gain financially from insisting that their cancer remedies were 'probatum'. In certain household receipt books, both topical and internal remedies promised to 'infallibly cure' cancer, to cure it 'tho it be eaten to the Ribbs' or was 'as bigg as a Goose Egg'.[69] These remedies were often similar – sometimes identical – to those contained in printed medical textbooks, yet their writers seem to have been far more optimistic about the likelihood of their producing a full and lasting cure. Once again, the reasons for this may be ones of how the disease was conceptualised and diagnosed. By and large, receipt book writers did not give cancer the special treatment it received in medical textbooks. Often conflated with other diseases such as King's-evil or scirrhus, there was no mention of cancer being 'evil' or 'rebellious' in these pages, of its peculiar appearance or rate of growth. Cancer appeared only as one more illness to be cured.

Across medical genres, physicians, apothecaries, empirics and practitioners of household physic offered a wide range of animal and vegetable remedies for cancerous tumours. Equally, they gave substantially different promises about how effective those remedies might be, based in large part on how narrowly 'cancer' was defined. While certain receipt book writers promised almost miraculous cures, others advised that 'we shall deale sufficiently in this case' if able to 'stop and hinder [tumours] growing and encreasing'.[70] Though their prognoses might differ, however, these remedies often shared a few key ingredients – some plants designed to strengthen and soothe, others which were extremely poisonous and animal ingredients from creatures felt to be literally or rhetorically aligned with cancer. This reflected the way in which cancer was conceptualised as both of and hostile to the body. In writing about these remedies there was less emphasis on rebalancing the whole body and much more on addressing the tumour as a hostile entity. Concomitant with this shift was a move away from self-prescribed and domestic physic toward professional intervention, and an increased emphasis on the fragile reputations of those who provided such intervention.

5.3   'Extreame remedies are to be used, against extreame diseases': pharmaceutical caustics and the first chemotherapies[71]

In printed medical textbooks, practitioners repeatedly emphasised the double bind which they felt that cancer presented. They widely maintained the conviction that harsh remedies exacerbated cancers. However, they often added to that conviction another, proven by bitter experience – that gentle remedies failed to touch the disease at all. This conundrum was nothing new to writings on cancer, yet it persisted over the entire early modern period. In 1571, for example, a translation of the work of fifteenth-century Italian surgeon Giovannida Vigo explained that remedies with 'a weake and feeble power ... worke no effect (as Galen saith) but are easily overcome'.[72] However, 'strong and mightie' medicines made the cancer-causing humours 'more obstinate and more hard to be dissolved and discussed'.[73] Over 120 years after this publication, a translated text by the French physician and surgeon Paul Dubé made an almost identical argument, asserting that cancer possessed 'so odd a Nature, that it does not hearken to gentle Remedies, and grows worse by the use of violent ones', while Culpeper's immensely popular 1651 *A Directory for Midwives* similarly complained that 'mild Medicines are not felt, and strong, exasperate'.[74] Each of these writers construed cancer in anthropomorphic terms, as having a will somehow independent of the sufferer. Cancer, they agreed, was resistant, stubborn and exasperating for medical practitioners, to say nothing of their patients.[75] This presented a serious problem. As Paré asked: 'seeing it refuseth gentle medicins ... and is not to be cured, but with strong medicins: which neverthelesse make it worse & more fretting, is [it] not to be deemed incurable?'[76]

For many medical practitioners, the answer to Paré's question was a simple 'yes', and they advised that treatment should be restricted to palliative care, in order to spare the patient further suffering. For others, however, this double bind did not signal the end of all curative efforts. If cancer ignored gentle remedies and reacted against stronger ones, the solution was to employ an arsenal of the era's most powerful medicaments, to deal a blow the disease could not resist. Ideas about what kind of substance might be best used to this end naturally varied widely.

In many cases, it was a matter of adjusting so-called gentle remedies by adding components such as alum, a potassium compound.[77] Other medical practitioners and domestic healers left the composition of their remedies unchanged but applied them at extremes of temperature, usually very hot.[78] The most notorious strong remedies, however, were those which were intrinsically and powerfully toxic. Providing the focus for the remainder of this chapter, they are perhaps the first recognisable chemotherapies – arsenic and mercury.

The *Oxford English Dictionary* defines chemotherapy simply as 'treatment with specific chemical agents or drugs', but the word has become synonymous in the past 50 years with a particular kind of pharmacy which visibly poisons the body in order to kill a cancer therein.[79] This rationale – poison against poison – was also at work in the early modern use of heavy metal and metalloid treatments for cancer. Medical texts of various kinds show that mercury and, to a lesser extent, arsenic were employed throughout the early modern period, primarily by physicians, but also occasionally by domestic receipt book writers or itinerant medicine-sellers. Those using mercury, for example, included noblewoman Elizabeth Grey, who used 'four pennyworth' of the substance in her recipe 'To make a Strong water good for a Canker, or any old Sore, or to eat any lump of flesh that growth [*sic*]'.[80] Practitioners varied in their explanations of just how mercury could remedy cancers. In therapies for venereal pox, it had long been accepted that the profuse sweating and salivation caused by mercury helped to expel bad humours from the body.[81] In texts on cancer medicine, however, this logic was less evident, and there was more emphasis on how the substance acted on the tumour or ulcer itself. In 1684, a translated work by the Swiss physician Théophile Bonet proposed that 'Leaden Plates smeared with Quick Silver, are a kind of Alexipharmack [antidote to poison], whereby the evil disposition of Malignant Ulcers is subdued and spent'.[82] At other points in mercury's long therapeutic history, medical writers recognised the toxicity of the metal as intrinsic to its effectiveness. In 1571, for example, Vigo extolled the virtues of mercury as not only a cancer cure in itself, but also an 'incredible' painless way to kill off any remaining 'superfluous' flesh left after cutting away a tumour.[83] In both models, the virtue of mercury in cancer cures was that it was powerful enough to 'subdue' the normally rebellious disease. Cancer ate the flesh; mercury, too, was capable of 'eating' unwanted or 'superfluous' flesh, demonstrating that it could match the strength of a malignant tumour.

Unlike mercury, arsenic powders and ointments were used more exclusively by professional medical practitioners and are not mentioned in household receipt books. Nonetheless, this too was a substance with a long therapeutic heritage. For instance, *The Surgery of Theodoric, c.*1267, recommended 'arsenic sublimate' as a way to mortify cancerous flesh so that it could be sloughed off.[84] Medical practitioners of the seventeenth century treated the substance in much the same way, with Culpeper, Riverius and Read among those authors who included arsenic, often in a 'sublimed' (washed) form, in their published cancer remedies.[85] On the continent, Anne of Austria (1601–66), who underwent various gruesome procedures in the hopes of curing her breast cancer, was treated with arsenic regularly between August 1665 and January 1666, with physicians applying arsenic to mortify the flesh and then cutting it away.[86] Across the late sixteenth, seventeenth and eighteenth centuries, the appeal of arsenic seems, like mercury, to have centred on its 'eating' qualities, which matched those of the cancerous tumour. Read's 1650 *Workes* recorded that 'superficiall' tumours could be 'eaten out' with arsenic, while in 1597, Lowe noted that arsenic possessed 'force to consume the evill humor'.[87]

Mercury and arsenic were material actors of such force that they, unlike diets, purges and herbal drinks, seemed able to match the ferocity of the cancerous tumour. Their ability to consume flesh mimicked that of the disease to be overcome, promising to expel cancer from the body in much the same way that medicines of worms and crabs seemed to work: 'like against like'. Since, unlike many medicinal ingredients, arsenic and mercury could not be used as foodstuffs, one might also view them as having had an additional psychosomatic force. These substances were firmly stamped 'medicine', and appear to have been well known as among the strongest remedies to be had. With this potency, however, came some drastic and dangerous side-effects. For every writer who recommended mercury and arsenic there were several more who warned in vehement terms that these substances were dangerous to the practitioner's reputation and to the patient's life. The unpleasant effects of mercury in particular were common knowledge among medical practitioners and many lay people, even cropping up in dramatic works such as Shakespeare's *Hamlet* (1.5.66–75).[88] Believed to act as a powerful purgative, 'salivation' treatment was associated with a raft of physical and neurological disorders, as described by Siena in his study of London's 'foul wards'. As well as excessive salivation, which was supposedly beneficial, '[t]he toxicity of ... prolonged regimens of a heavy metal usually produced dreadful side effects. Patients frequently suffered internal pain, intense nauseam and permanent damage to their mouths including loss of teeth, gum damage, and the complete loss of the uvula'.[89]

It is unclear whether mercury treatment for cancer was as prolonged as that for pox, which could last five weeks or more.[90] It seems unlikely that patients with advanced cancers could have survived such a regimen. Nonetheless, even those who advocated mercury treatment admitted that the substance could be dangerous. Reporting on the case of a woman with breast cancer, Bonet recorded:

> The Physician that was consulted ordered a Plate of Lead to be applied, and every other day to smear it lightly with quick silver ... But through the carelessness of those that lookt after her, the Plates did more harm than good. In the mean time the Canker encreased, and came to Suppuration; therefore the use of the Plate was laid aside. The Swelling broke of itself, and her torments ceased a little; but by and by they returned more violent and pungitive, the Canker encreaseing in all its dimensions. It deserves admiration, that the Mercury which was formerly imbibed from the Plate, should drop so visibly, and in a pretty quantity out of the *Carcinoma*, which shaded the adjacent parts with its shining, nay, and sweat at the shoulders through the whole skin.[91]

In this case, the parallel properties of mercury and cancer turned against the patient. The 'eating' mercury failed to consume the cancer, but led to 'suppuration' and allowed the cancer to keep growing, setting the stage for the onset of a cancerous ulcer. Worse still, mercury then spread through the body, becoming visible at its surface in a way which brings to mind cancer's much-discussed propensity to remain 'secret' before suddenly 'discovering' itself. Altogether, the account demonstrated vividly the dangers attendant on introducing a substance to the body which might exceed the practitioner's control.

If mercury was a source of anxiety, arsenic caused outright panic among some early modern medical practitioners. Unlike mercury, there is little evidence for the substance having been notorious in popular culture, but medical writers recorded numerous instances of arsenic's fatal side-effects, often in lurid terms.[92] The expanded 1712 edition of William Beckett's *New Discoveries* tempered acknowledgement of the popularity of arsenic among some medical practitioners with a striking warning from personal experience:

> This Powder [of Arsenic] I apply'd to a *Cancerated* Breast of a Woman, under thirty Years of Age, after having made a Sore by applying one of the milder *Causticks*, the night it was made use of, it caus'd a great

deal of Pain, and the next Day, the Breast appear'd very much tume-
fied and inflam'd...in short for fifteen Days she was not free from
pain, she had a *Fever*, was attended with frequent *Vomitings, Faintings*,
and several other Disorders. I cou'd afford her but very little Relief
by intervals...nor was it in my Power to remove the Dressing, it
adhaered so fast to the Sore. There was a discharge of a bloody serous
Juice for twelve Days in a moderate quantity, after which the matter
thicken'd, and it began to smell somewhat offensive, at the end of
fifteen Days the Dressing drop'd off, and with it came away about two
Ounces of the *cancerous Mass*. The Reader may easily imagine that
making so small a progress in such a time, and that at the Expence
of so much Pain, I cou'd easily prevail with myself to desist from the
undertaking, for the second Application wou'd have been attended as
the first, which to any Person that entertains such a concern for his
Patients as he ought to do, must be very fatiguing.[93]

In this account the ability of arsenic to redouble the disease's original
symptoms was again apparent, with the medicine producing pain,
'serous Juice' and stench, as well as 'adhering' to the body in much the
same way as the obstinate, crab-like tumour. Even more distressing for
Beckett was the immense pain to which the therapy put his patient.
Not only 'fatiguing', this effect was sufficient for Beckett to abandon
the use of arsenic altogether, stating that 'we can't say, but there are
many Cancers that may be cur'd by Causticks, but the Person that is
to undergo it, may very well answer...*The Preservation of Life would be
too dear bought at the Price of so much Pain*'.[94] Beckett's account was far
from isolated. Only a decade earlier, Browne had likewise attested that
arsenic could cause 'Faintings, Swooning, Fever, Madness' by sending
forth 'malign...vapours' to the heart or brain even when it was applied
to the arms or legs.[95] Like mercury, arsenic threatened to break its
bounds, taking over the body in the same manner as the cancer itself. At
the mildest end of the cancer treatment spectrum, dieting and purging
had treated a patient's whole body and often their mind. Arsenic and
mercury now targeted the cancer in isolation. They were clearly far more
potent treatments, but they could kill the sufferer before they quelled
the disease.

Browne's warning of the 'great injury' wrought by arsenic, as well
as Beckett's 'fatigue' at witnessing his patient's suffering, points to an
uncomfortable awareness among medical practitioners of the risks to
which they exposed themselves as well as their patients when they
administered dangerous remedies. In the 1684 translation of Bonet's

Guide to the Practical Physician, for instance, the fate of a rival practitioner's patient was described thus:

> I have observed [septics], especially Arsenick, and sublimate in a greater quantity, and not tamed, applied to Ulcers near the heart, as to a Cancer in the breast, that they once carried off a Woman in 6 days: About three hours after the Powder was strewed on her Breast, she just as if she had swallowed it, was taken with a Shivering, then with a Vomiting, and frequent Faintings, with a languid Pulse; which symptomes, encreasing by degrees, her extreme parts growing cold, and her Face and whole Body swelling beyond measure, she was miserably murthered.[96]

The 'murther' Bonet described demonstrated forcefully the moral predicament facing those who administered arsenic or similarly toxic medicines. Although the substance held some potential to cure an otherwise fatal disease, it also presented an imminent danger to the life of the patient. Moreover, Bonet went on to name four practitioners who used arsenic in their cancer medicines, and were therefore to be avoided.[97] Being associated with arsenic treatments could therefore be economically as well as morally dangerous in a marketplace where the consumer – at least, if London-dwelling and affluent – had various practitioners seeking their custom, and in which physicians accused of malpractice could find themselves fined or even imprisoned.[98] In this climate, why did both patients and practitioners continue to pursue such dangerous and toxic treatments?

The answer to this question may partly be found in the diaries of Reverend John Ward, with which this book opened. Recording a physician's prescription of 'desperate' cordials to a seriously ill patient, he recalled this maxim: 'With it they may die, but without it they will die'.[99] One must remember in the case of dangerous medicines (and in the following chapter, of surgery), with what stakes patients were gambling. Cancer, as we have seen, was firmly established in the popular consciousness as a cruel and fatal disease, such that in many cases patients must have viewed the use of extreme pharmaceuticals as risking possible swift death, with the chance of cure, against certain, perhaps slow death, with no chance of reprieve. In addition, mercury and arsenic, hazardous as they undoubtedly were, may still have seemed a favourable alternative to the other means by which a cancer could be 'consumed': namely, cautery with burning irons, or surgery. Patients making the seemingly extraordinary choice to be voluntarily poisoned by mercury and arsenic may have exhausted more gentle means, and experienced their options as a matter of choosing the lesser of two evils. Furthermore, scholars of early modern medicine suggest that the painfulness of certain therapies may actually have been taken as a marker of their effectiveness. Michael Schoenfeldt, for instance, emphasises the centrality of pain to early modern experience, including the belief in pain as a form of divine punishment.[100] Writing on medical metaphors in moral theology, David Harley similarly notes that whilst believers were not discouraged from seeking medical relief from illness, painful medical treatments were frequently compared to confession or repentance.[101] As with the administration of purges or burdensome regimen, going through painful remedies might be construed as penance for one's inevitably sinful nature, or as a kind of labour by which to expel the disease from the body. As Porter argues, '[T]he ferocious painfulness of a treatment might even work in its favour – the earnest of its efficiency lay in its bite or sting'.[102]

The factors which made medical practitioners stake their reputations, and thus their livelihoods, on arsenic and mercury as cancer cures are less obvious. Although they had much less to lose than their patients, those administering extreme pharmaceutical remedies also had less to gain. There is little to suggest that physicians or medicine-sellers built lucrative commercial reputations based on curing cancers with these chemicals. On the contrary, they were likely to be decried by their fellows. One component in encouraging medical practitioners to use arsenic and mercury despite the risks may well have been compassion for the suffering of their patients. Surgeons frequently stated that they were induced to perform dangerous operations by the pleas of the sick party, and it seems reasonable to suppose that physicians were subject to the same pressures. In addition, of course, they must have been aware that should they refuse to administer certain therapies, a patient might simply go elsewhere. The language in which practitioners describe their use of mercury and arsenic may also offer clues as to why they persisted in this dangerous course. Where dietary cures involved the treatment of the whole body, and the active participation of the patient, cures by arsenic and mercury were often framed as attacking the cancer in isolation from the rest of the body. For example, describing the use of arsenic powder by 'Fuchsius', Browne noted that

> he applied [arsenic powder], upon which, if the Cancer did not grow more angry the 3d day after, he declared the Cancer curable; and if it grew better, the Powder was to be kept on for 30 days, in which time

it would be eradicated from the very roots, and they fall off of themselves; and if any part of them did continue adhering, he usually cut it off with his Knife.[103]

Despite the considerable pain this must have caused Fuchsius's patients, they were virtually invisible in this account, having neither voices to assent or protest, nor any discernible role in their own treatment and recovery. Tellingly, it was the cancer, not the patient, which was deemed 'curable', but which could become anthropomorphically 'angry' upon application of the powder. The ferocity of the arsenic powder, which caused tumours to 'fall off', is equally telling. Chosen because its 'eating' properties matched that of the malignant cancer, one may see the curative agent in these accounts as an extension of the practitioner's own strength. Indeed, even as he warned of the difficulty of employing chemical caustics, Wiseman reported that 'they do your work in less than an hour'.[104] Just as cancerous disease was construed in zoomorphic and anthropomorphic terms, portrayals of potent cancer remedies aligned the substances with a sentient agent – in this case, the ministering physician.

Arsenic and mercury were not the most commonplace remedies for cancer in the early modern period. In the imaginations of medical writers and their audiences, however, they loomed large, as much for their dangerous side-effects as for their potential to cure. Many medical practitioners, and presumably their patients, were nervous about using substances which caused such drastic collateral damage. For some, the risks seemed unjustifiable, and edged the physician or apothecary over the tenuous boundary between healing and harming, at which point they became no better than 'murtherers'. However, many others continued to employ heavy metal and metalloid ingredients in their cancer remedies. The appeal of such potent substances was of a piece with the discursive estrangement of patients from their alien, invasive tumours. In the battle against cruel, 'rebellious' cancers, arsenic and mercury were powerful armaments.

Conclusion

In the diverse accounts of early modern cancer medicines, patients' voices are conspicuously absent. An appropriate postscript to this chapter is therefore provided by the poignant but frustratingly incomplete record of one woman's experiences with cancer physic, as told by her uncle Henry More, in a series of letters written to Lady Anne

Conway between July 1674 and January 1676. More and Conway were in correspondence on a number of matters. However, More sought Conway's advice in particular regarding his niece 'Mrs Ladd', who was suffering from breast cancer, because of Conway's close acquaintance with 'Monsieur Van Helmont' (son of Jean Baptiste Van Helmont). Via Conway, Van Helmont sent prescriptions to More and Mrs Ladd which are not detailed in the surviving documents, but, judging by their effects, contained some potent chemical and organic components.

Beginning optimistically, More's letters described his hopes of a cure for his niece, and told how he had informed her of Van Helmont's 'fame' in Europe, hoping that 'it may contribute to the efficacy of the medicine'.[105] Within a month, however, Mrs Ladd began to experience the side-effects of her treatment. Her physician informed More that the medicine

> produced no alteration in her till the Sunday following, she has been these three dayes ill at her stomack, hot and thirsty, with frequent shootings in her breast, and not only on the cancer'd part, but likewise round about it there are many little angry pustulats, first red, afterwards maturated on their heads.[106]

Nonetheless, seeing some softening of the tumour, Mrs Ladd was persuaded to carry on. Over the coming months, she repeatedly complained of pain and fever, sometimes declining to take the remedy, then consenting to its use once more. In December 1674, More wrote to Conway:

> I hear from Grantham also that my Neece make use againe of the Plaisters ... But in a letter December 3 my Nephew writes thus, Though my sister be advised from ragley to proceed in the use of her plaister, yet I doubt she needs further advise what to doe; for besides the paine and disturbance it hath given her, it has much encreased the bignesse of the soar which is all in such fretting distemper. This makes me tell Dr Clark that he must judge upon the spot. And I beleeve he does not deele so openly with me as he should, out of a nicenesse to displease me, Because of my great opinion I have expressed of Monsieur Vanhelmont. So that I am something at a losse what to doe in the case and dare over sensibly presse the use of the plaister upon my Neece. For feare of the worst.[107]

Shortly after, he reassured his correspondent that 'what ever it be I shall account myself much obliged to Monsieur Vanhelmont for his good

will. He did not pretend to ascertaine the cure at first. But seeing by
this Medicine he had cured this kind of disease, I could not but take
the boldnesse to desire him to try the successe of it on my Relation'.[108]
Mrs Ladd then disappeared from More's letters. When she reappeared in
March 1676, it was for More to inform Conway that he was journeying
to his niece's deathbed.[109]

More's letters provide the closest thing to a patient's account of
pharmaceutical cancer treatments in this period and are a salutary
reminder of the real sufferings behind textual representations of medicine. They also show the shifting relationships which cancer sufferers
had to their physicians and other medical practitioners at this time.
More, Mrs Ladd, and those around her were seemingly caught between
a desire to acquiesce to the 'famous' physician, and conviction in their
own observations, that the cancer was being exacerbated by his so-called
remedies. Accordingly, the story of non-surgical treatments for cancer is
a complex one. More than any other facet of the diagnosis and treatment
of cancer, one might expect non-surgical treatments to show substantial change over time, influenced by the much-discussed rise of iatro-
chemistry in the later seventeenth century. The sources, however, give a
more nuanced account. Medical practitioners may or may not have used
mercury and arsenic with increasing frequency over time. Accounting
for the bias toward the later part of the early modern period created by
material factors – namely, the increased number of texts produced, and
hence available to us today – it is difficult to see any conclusive evidence
of a move toward these kinds of remedies. Certainly, neither medical
practitioners nor patients of the seventeenth and early eighteenth
centuries were prepared to abandon gentler prescriptions for regimen
and medicines explicitly aimed at correcting the humours. Furthermore,
the inclusion of mercury and arsenic cures in medieval texts prohibits
us from imagining these substances as having been 'discovered' by
Renaissance physicians, or taken up as a direct result of the rise of iatrochemical medical models.

What can be traced, however, are smaller-scale shifts in rhetoric and
practice identifiable with the changing ambitions of both patients and
those who treated them. Cancer was by no means the only intractable,
fatal disease of the early modern period. It was, however, among those
most vividly imagined in zoomorphic and anthropomorphic terms,
a disease which was, as we have seen, both of the body and alien to
it, which seemed purposely malign, evil and rebellious. As patients
became increasingly pained and frightened by progressive cancers, it
is little wonder that they sought remedies of increasing strength and

complexity, tolerating the discomfort they evinced and even taking their suffering as signs of the treatment's efficacy. Significantly, one can also see in these discourses how, through the process of increasingly desperate cures, patients relinquished – or medical practitioners appropriated – authority over their bodies and what happened to them, a move concomitant with the construction of cancers as 'independent' from cancer patients. Adjustments to diet and regimen, the first recourse for most cancer patients, closely involved the sufferer in their cure, and were readily explicable in terms of a holistic humoral system. A little further along the treatment spectrum, medicines containing herbal and animal ingredients were increasingly targeted at the tumour, rather than the individual patient, allowing medicine producers to peddle 'one-size-fits-all' cures. The rhetoric which accompanied description of the harshest non-surgical cures, however, shows how the therapeutic landscape was shaped as much by language as by science or economics. With cancer constructed as evil, medical practitioners cast their own attempts to cure the disease as a battle of medical knowledge against a discrete, zoomorphic enemy. The diminished figure of the patient, therefore, made room for the confrontation between cancer and physician to be writ twice as large; a trend which would continue to develop in the agonising procedures of cancer surgery.

6

'Cannot You Use a Loving Violence?': Cancer Surgery

In fury Quintianus ordered them to torture her by crushing her breasts, and when she had suffered in this way for many hours, he finally ordered that her breasts be cut off. 'Impious, cruel, odious tyrant!' Agatha cried. 'How could you do this? Are you not ashamed to take from a woman what your own mother gave you to suck? No matter: I have other breasts you cannot harm, breasts that give spiritual nourishment to all my senses, and them I dedicated long, long ago to God'.[1]

Saint Agatha, an early Christian martyr, was popularly believed to have had her breasts removed as a method of torture. The young Christian, living in ancient Sicily around 231 AD, had caught the eye of the 'idolatrous' governor Quintianus, who, angered by her rejection of his sexual advances, had her arrested for her faith and imprisoned in the house of Aphrodisia, a prostitute who attempted to persuade Agatha to welcome Quintianus's attentions.[2] Finding that she remained unmoved, Quintianus ordered Agatha to be tortured by having her breasts mutilated and cut off. Then, infuriated by the composure with which Agatha bore this punishment, he had her thrown into a dungeon and left to die. Quintianus's final revenge, however, was futile, since Saint Peter appeared to the stricken Christian and restored her breasts. She died after later being rolled on hot coals, an avowed martyr of the faith.

Agatha's story struck a chord in early modern society. She appeared everywhere from Greek and Latin martyrologies to classical poetry and the works of the early Baroque artists, who depicted her undergoing torture or serenely carrying her severed breasts on a platter (Figures 6.1 and 6.2).[3] Her story was recounted at length in the influential medieval martyrology *The Golden Legend*, a text that was 'without doubt one of

Figure 6.1 Lorenzo Lippi, *Saint Agatha*, 1638–44, oil on canvas, 75.7 × 64.1 cm (29 13/16 × 25 1/4 in.). Courtesy of Blanton Museum of Art, The University of Texas at Austin, The Suida-Manning Collection, 1999 (369.1999). Photo credit: Rick Hall. This image is open access under a CC-BY 3.0 licence.

the most widely disseminated books through Europe from ... 1266 until the end of the Middle Ages'.[4] Most intriguingly, she was, argues Liana de Girolama Cheney, at the centre of resurgence in 'porno-violent hagiography' near the end of the fifteenth century which continued into the sixteenth and seventeenth centuries and was 'augmented by the writings of anatomical science and medical texts'.[5]

Figure 6.2 Sebastiano del Piombo, *The Martyrdom of Saint Agatha* (1520). Courtesy of Polo Museale, Firenze, the Vittoria della Rovere collection.

As the patron saint of breast cancer patients, Agatha later gained an associate of sorts. Born in 1265, Saint Peregrine was the youngest member of a wealthy Italian family active in the antipapal movement of that period.[6] Upon a visit of the papal ambassador to his locale, it was said that Peregrine joined others in harassing the ambassador and struck him in the face. The ambassador promptly forgave Peregrine and prayed for him, upon which the young man was so moved that he converted to Catholicism and joined the Order of Servants at Siena. Following many years of an ascetic lifestyle in which he never sat or lay down, Peregrine developed a leg ulcer which was pronounced cancerous, and he was told that amputation was the only cure. Peregrine spent the night before the planned operation praying in the chapel, and on falling asleep, dreamed that Christ reached out and touched his leg. Upon waking, the monk found that his leg had healed, and went on to thrive into old age. While his story was a medieval one, Peregrine's beatification took place in 1609. His corpse was repeatedly dug up, and found to be uncorrupted, throughout the seventeenth century, and he was canonized by Pope Benedict XIII in 1726. In early modern Europe, therefore, there was a great deal of interest in this cancer survivor – some of which, despite widespread anti-papist feeling, must have crossed the seas to England.

What did Peregrine and Agatha have in common, and why did they both become prominent during the early modern period as icons for those facing cancer, despite their radically different experiences? The link between the two figures seems to have been amputation: facing it, suffering it, avoiding it or recovering from it. Agatha remained serene throughout a stylized rendition of a double mastectomy. Peregrine's reprieve from surgery appeared as a powerful example of wish fulfilment. By enduring or avoiding the knife, the two saints reflected the worst fears and most ardent fascinations of their audiences. It is with Peregrine and Agatha in mind, therefore, that this chapter examines representations of surgery to analyse what they reveal about early modern attitudes to cancer, cancer sufferers and medical practitioners. What was cancer surgery? How did it relate to perceptions of cancer, or of the nature of the gendered body? And why would anybody consent to such a 'frightful' course?

My analysis of cancer surgeries, surgeons and patients in this period builds on the contention of Chapter 5: that constructions of cancer as alien to the body encouraged an adversarial therapeutic approach, in which the patient's individuality and subjectivity were often eclipsed. For surgeons, as for physicians, it seems that the intractable, 'rebellious' nature of cancerous disease was felt to justify, and even to demand, the use of radical therapies despite their inherent risk to the patient. For surgeons, however, I argue that the issues raised by dangerous pharmaceutical treatments were amplified. Cancer surgeries – in particular, mastectomies – were among the most dangerous and invasive of the era's medical procedures. Temptingly, they offered a means to remove the perceived interloper from the body, a last resort for patients who believed that they otherwise faced certain death.[7] However, while this course offered chances for glory, it also supplied disruptions to the narrative of medical progress. Surgeons who carried out cancer operations might find themselves denounced as reckless butchers, or frustrated in their curative efforts. In short, stories of cancer surgery display all the potential and problems of a discourse which sought to divorce patients from their misbehaving bodies.

In the scholarly literature, surgery for cancer has been recognised as an ancient but rare phenomenon. Numerous authors have recognised descriptions of surgical excision of tumours dating back to ancient Egypt and the Edwin Smith papyrus.[8] For the medieval period, Luke Demaitre notes that several authors listed surgery as among the possible cures for cancer, though they counselled readers to avoid this course.[9] Marie-Christine Pouchelle's *The Body and Surgery in the Middle Ages* also identifies Henri de Mondeville, an eminent fourteenth-century surgeon, as having performed a variety of operations to remove cancerous tumours and ulcers.[10] Writing on the more recent past, Marjo Kaartinen argues that the mastectomies for cancer were relatively common during the eighteenth century, and finds the latter half of the century to have been marked by the development of 'radical' forms of mastectomy in which much underlying muscle was removed.[11]

My own analysis has been influenced by a growing literature on the semiotics and practice of Renaissance surgery, much of which contradicts stereotypes of the 'swashbuckling "sawbones"' heedlessly hacking off limbs and pulling teeth.[12] Lynda Ellen Stephenson Payne's *With Words and Knives: Learning Medical Dispassion in Early Modern England*, for example, provides a thoughtful look at surgeons' attitudes toward patient suffering, reading between the lines of texts which take a brutal approach to those under the knife, and demonstrating that many surgeons were keenly aware of the pain they inflicted.[13] Taking a broader view of surgical practice, works by Andrew Wear and Philip K. Wilson describe a medical landscape in which surgeons formed an increasingly professionalised and learned body, with ambitions toward the same prestige and rewards enjoyed by members of the Royal College of Physicians.[14] With the translation of many classical anatomical texts into English, an increasing number of surgeons possessed scholarly credentials to match their substantial practical training, and 'English reformers of surgery', argues Wear, 'stressed with great unanimity that both groups [physicians and surgeons] had much in common in terms of medical theory and practice'.[15] Moreover, Wear finds surgery to have been a more dynamic field than physic, open to innovation in procedures and instruments and with 'a craft emphasis on practicality, dexterity and the value of experience'.[16] From 1684 onward, surgeons repeatedly applied for their craft to be divorced from that of the barbers with whom they shared a College, a wish finally granted in 1745.[17] While Wear and Wilson have illuminated surgeons' ambitions for their profession, work on perceptions of surgery among non-medical audiences has been less forthcoming. As I will discuss later in this chapter, however, several scholars investigating the representation of early modern torture, vivisection and anatomy have noted that these crafts were often compared with surgery, such that the surgeon's status as a preserver of life was often tenuous.[18] This aspect of the semiotics of surgery, especially invasive surgery, begs further study, and my examination of the possible affiliation of cancer surgery with these cruel and violent trades aims to contribute to that broader discussion.

This chapter is divided into four sections, focussing first on questions of why and how cancer surgeries were undertaken, and later on the difficulties of representing these operations in medical writings. In the first section, I address two obvious questions – whether cancer operations were taking place, and why patients might consent to them. The second section then looks at methods for some of the most common procedures. Section three considers what motivated some surgeons to carry out cancer operations, and how that motivating narrative came under threat from fellow medical practitioners. Finally, I examine how issues of gender and power were treated in accounts of surgery from the operators themselves.

6.1   'But is there no other Way, but this frightful one?'[19] Facing cancer surgery

Any examination of surgery in the early modern period – an era before antiseptics, antibiotics or anaesthesia – must begin with several obvious questions. Did cancer surgeries actually take place during this period? If so, then why? That is, why would anybody consent to have their body cut into, even to have parts of their body amputated, when doing so ensured agony and potentially death? In this section, I contend that cancer surgeries were an established feature of the early modern medical landscape, and that patients' decisions to undergo these procedures were based on personal experiences of suffering as well as popular beliefs about cancerous disease.

Accurately quantifying cancer surgeries is an impossible task. Most of the surgical practice actually taking place in this period was never recorded, much less preserved for modern readers, and medical textbooks often provided instructions for an operation without indicating whether the writer had actually carried out that procedure, or how often. In her study of breast cancer, Kaartinen suggests that surgery 'became more common' from the late seventeenth century onward, and provides numerous examples of mastectomy from the mid-late eighteenth century.[20] In the period 1580–1720, however, the picture is less clear. Cancer operations seemingly remained uncommon, and, as I shall discuss, many medical practitioners and patients refused to countenance the procedure, for a variety of reasons. Nonetheless, anecdotal evidence suggests that cancer surgeries were an established feature of the early modern medical landscape. For instance, in May 1665, Samuel Pepys remarked upon the mastectomy of his 'poor aunt James' with sympathy, but without much surprise.[21] Some medical textbooks, most notably

Wiseman's *Several Chirurgical Treatises*, gave many examples of surgeries the authors had carried out, including dates, locations and names. Most tellingly, numerous newspapers carried advertisements indicating that cancer surgeries were taking place on an infrequent but steady basis during the early eighteenth century. On 8 February 1728, for instance, an announcement in the *London Evening Post* reported that 'the lady of Sir Challenor Ogle' had undergone an operation to remove a cancer in her breast, 'and there is great Hopes of her Recovery'.[22] Newspapers' obituary pages also indicated the prevalence of cancer surgery, albeit in unhappier terms: multiple listings record the deaths of cancer patients during or following operations, most often mastectomies.[23]

Clearly, some cancer sufferers did opt for surgery, despite its evident perils. Moreover, descriptions of surgery, as seen later, indicate that they were often doing so in a premeditated and considered manner, when it did not seem that their disease was immediately about to kill them. This fact makes cancer surgery particularly interesting. Other amputative or invasive procedures described for the same period tended to take place after accidents or on the battlefield, with death otherwise imminent. The most notable exception to this rule, lithotomy, was usually completed in a matter of minutes, whereas cancer surgeries could take hours or even days.[24] Cancer operations, almost uniquely, entailed a patient agreeing, in advance, to lay down their more or less functional body for prolonged cutting and burning knowing that they might never get back up. Estimating just how many such patients did get back up is a fraught undertaking. In his study of the work of surgeon Daniel Turner, however, Wilson has found that tumour patients fared worst of all those whom Turner attended, with 28.9 per cent dying in the practitioner's care.[25] Turner was by all accounts a skilful surgeon, and Wilson's analysis does not specify how many of these patients underwent mastectomies or amputations versus the number treated with more conservative lumpectomy or cautery. It thus seems clear that many, perhaps even most, patients undergoing substantial cancer surgeries would die during or soon after their treatment.

Given these appalling odds, what made cancer patients agree to, or even seek out, a surgical cure? Firstly, patients experienced an increasingly poor quality of life as their illnesses progressed, and grasped at any chance, however remote, to end their pain. Secondly, the formulation of cancer as a rebellious, semi-sentient, unstoppably malignant disease impelled patients to remove these seemingly alien growths from their bodies before they took over. Evidence for the first of these considerations was stressed in texts discussing cancer surgery, where surgeons sought, as I shall discuss, to justify their involvement in such risky cures. Poignant accounts from these surgeons' case records depicted patients often unable to lead any semblance of a normal life, in constant pain and suffering social isolation as a result of their illness's appearance and putrefactive stench. Wiseman, for example, described the following encounter with a patient suffering with mouth cancer:

> Coming to the Patient with the [palliative] Prescriptions, he asked what way we had designed to cure him. After some pause (for we, having no hopes of curing him, had not discoursed of that,) Sir *Fra. Pr* answered, the attempt of Cure in such Ulcers had been always unsuccessful and extream painful...and thereby the Disease hath been exasperated, and the Life of the Patient shortned. The same was affirmed by us all. The Patient replied, God's will be done. I pray go and consider of the way: for I had rather die than live thus.[26]

The patient in this account suffered from a tumour and ulcer that had caused most of his teeth to fall out, and had spread from his jaw to his cheek and the roof of his mouth. Daily life – eating, drinking, talking and sleeping – must have been painful and laborious in the extreme, and it was this loss of function, even more than the attendant pain, that Wiseman later described as the motivation for patients putting 'to trial' a cure 'by Knife or Fire'.[27] Given that patients with quite minor tumours were often tempted to undergo surgery, he asked:

> How much more then shall these poor creatures, who have Cancers over-spreading their Mouth, eating and gnawing the Flesh, Nerves and Bones? Who, besides the danger they are in every minute of being choaked with a fierce Catarrh, do suffer hunger and thirst; and if they can swallow Broth, Caudle or Drink, yet is it with an unsavoury tast...and their Spirits are infected with the stink, whence Fainting frequently happens; Sleep is a stranger to their eyes, their Slumber very troublesome, and Death is only their desire. At such a time as this it is not to be wondred if they try a doubtful Remedy, though painful.[28]

Pain and debility were in themselves strong motivators for undergoing surgery. In the case of cancer, however, those pains were felt all the more keenly in light of their relation to the fearsome 'nature' of the disease. In opposition to surgery, as to aggressive pharmaceutical treatments, practitioners repeatedly cited Hippocrates' aphorism 6.38: 'Occult cancers ought not to be cured; for they that are cured die soon, whereas they that are not cured live longer'.[29] However, as the inclusion of cancer remedies in many of those same texts testifies, many patients could not be satisfied with such measures. Moreover, the construction of cancer in zoomorphic terms, with repeated emphasis on its malign, rebellious and 'cruel' characteristics, framed the ideal response to the malady as its physical *removal* from the body, a desire that seemed realisable only by surgery. As Théophile Bonet put it, '[Y]ou must try even with danger to cure a Disease, that would certainly kill'.[30] Although many writers gave examples of patients who lived with tumours until their death from some other cause, for those experiencing bodily 'invasion' by cancerous tumours, these examples paled in comparison to the tales of cancer's malignancy reinscribed by both medical and popular literature. In this climate of fear, Dionis bluntly advised one patient that 'she had no other choice, but either that Operation [mastectomy] or Death'. 'She, *like all other Patients*', he recalled, 'preferring Life to the Loss of a Member, determin'd to undergo it'.[31]

Accounts of the circumstances which led cancer sufferers to consent to, or even demand, surgery offer a vivid picture of patient experiences of this disease. Whilst the noting of cancer operations in newspapers implies that these procedures were uncommon, the way in which they are presented nevertheless shows that they were an established treatment route for cancers, regardless of the risks they posed. The individual decisions which led to these operations – the extraordinary acts of consent to amputation and incision made by patients – were based on prolonged suffering and the belief that that suffering could be ended only by expelling the malign 'alien' from within. In these critical decision-making moments, the thoughts and feelings of the patient are, perhaps unsurprisingly, visible to a greater extent than anywhere else in the surgical process. Their experiences show poignantly the distress they experienced every day, and for most patients, this was the only stage at which their opinions about their surgery, good or bad, would be recorded. As I shall demonstrate, when they came under the knife, cancer sufferers' voices subsided, and they were presented – ideally at least – as passive, silent bodies.

6.2 Operational methods

Diseases which Medicines cure not, the Knife cureth; what the Knife cures not, Fire cureth; what the Fire cures not, they are to be esteemed incurable.[32]

Descriptions of what drove patients toward surgery usually foregrounded individual patients' suffering. When the decision was made, however, and the patient came under the knife, the emphasis of surgical texts changed drastically. As in this discussion of 'Knife' and 'Fire', by the German medical practitioner Johannes Scultetus, medical textbooks and casebooks shifted their focus from patients to bodies, and from bodies to tumours. This new perspective was centred on 'extirpating what is super-fluous', and there were diverse methods by which surgeons could do just that.[33] Some cancer operations were relatively minor, while others posed a serious risk to the patient's life. Some were the work of minutes, while others took days to complete, and they could be undertaken on parts as diverse as the eyes, breasts, face, legs, and scrotum.[34] This section identi-fies three main operations which constituted the vast majority of cancer surgeries, and which each showed relative homogeneity across the early modern period and the diverse locations in which they were performed. These paradigmatic cancer operations – ordered here in terms of their increasing invasiveness and dangerousness – were simple lumpectomies, facial surgeries, and mastectomies.

For any operation, certain preparations had to be made and precautions taken before the patient came under the knife. As Wiseman observed, operating in the spring or autumn was preferable, though not always possible.[35] In many cases, surgery represented the last resort in a course of treatment, so it was likely that the patient would already have been eating a prescribed diet and perhaps taking medicines aimed at redu-cing the tumour and strengthening the body. Where mitigating pain was concerned, Kaartinen argues that eighteenth-century surgeons often administered opiates and alcohol before a procedure. Although they showed concern for patients' pain, however, most accounts of cancer surgery prior to 1720 make no reference to any such ministrations. This might have been because surgeons were aware of the possible risks of overdose with opiates in particular: as I shall discuss, records of palliative care show that medical practitioners were happy to prescribe laudanum to patients who were clearly dying, often to help them sleep, but they were conscious of the medicine's potentially lethal side-effects. In addition, it was often necessary that the patient remain conscious so that the opera-tors could gauge his or her physical state. Sudden sensitivity to the knife might indicate that a surgeon had reached the bottom of a necrotic ulcer and touched living flesh; conversely, slipping into unconsciousness was a worrying sign of blood loss as well as a natural reaction to intense agony.

Tumours which appeared on the face, arms and legs often merited relatively minor surgeries (insomuch as any early modern surgery was 'minor') which were designed to bring the malady to a swift conclusion while minimising its physiological and social impact on the patient. As Alexander Read pointed out for 'apostems' (undifferentiated, generally benign, lumps), surgery might be preferable to some medicines, particularly caustics, in such cases: 'First, if Apostems be in the Face, to avoid the filthiness of the Scar, after the Curation. Secondly, in small Tumors: for so they will be the sooner whole'.[36] Philip K. Wilson and Olivia Weisser separately note 'the stigma of a marked body': namely, that marks or moles on the face were often taken as signs of bad luck, or worse, symptoms of venereal disease.[37] Patients might thus have been tempted to undergo this procedure even where tumours appeared slow growing or benign. Worried sufferers may also have been fearfully aware of cases in which facial tumours ulcerated and 'ate' through the cheeks, nostrils or eyelid.

In the best cases, excision of small tumours could provide a quick, if painful, resolution to the problem. Wiseman, for example, cited the example of 'A Man of about fifty years of age...with a hard unequall Tumour, of the bigness of a large Wall-nut, between the Coronal and Sagittal Suture'.[38] This tumour, Wiseman recalled, 'was at that time crusted over with a Scab, and seemed to be a milder sort of Cancer'.[39] Wiseman decided to operate:

Therefore providing Dressings ready, I made an Incision round it to the Scull; then raised it off with a *Spatula*, and permitting the bloud to flow a while, dressed it up with Astringents. The third day after I took off Dressings, and saw the Lips of the Wound well disposed, and the *Cranium* uncorrupted. I rasped it till the bloud appeared under it, then dressed up the Wound with Digestives...and after Digestion incarned and cicatrized it with as little difficulty, and dismissed him cured.[40]

Several factors contributed to this operation's success. The tumour was, as Wiseman noted, 'resting upon the *Cranium*', a hard base from which it could easily be separated. The lump was relatively small, and the patient was acquiescent to Wiseman's method, allowing him to apply medicines and cauterize the wound over several days. Wiseman's description, however, was atypical of the kinds of operation most frequently found in medical textbooks. Whether because they were felt not to merit recounting, or because they were rarely carried out, straightforward excisions of sub-dermal tumours were the exception rather than the rule. Most descriptions of cancer surgery on the face and limbs recorded rather more complicated procedures, often with less positive outcomes.

Despite the distinctive symptoms identified by various medical practitioners as signalling cancer, it is clear that many patients, particularly those travelling from the countryside to seek medical advice in the city, did not identify their tumours as cancerous until they reached an advanced stage. Furthermore, they were understandably reluctant to consent to surgery until it became clear that there was no other option. This state of affairs may explain why most of the facial cancer surgeries described in medical texts (and among cancer operations, facial surgeries far outstripped everything but mastectomies) tended to be lengthy, often complex affairs. Surgeons described operations for tumours which had spread over the face, often involving the gums, nasal cavities, eyelids and even the eye itself. For instance, in another of his many examples of the difficulties of cancer surgery, Wiseman recounted the case of a 'military Captain' whose initially minor mouth cancer had spread to include the salivary glands, both '*Maxilla*' (bones of the upper palate), the lower lip, the gums (causing some teeth to fall out) and some glands under the jaw.[41] On consulting Wiseman, the patient was informed that his tumour was cancerous, and resolved to have it removed by Wiseman with the help of fellow practitioners Thomas Cox, Walter Needham and 'Mr. *Gosling*'.[42] Wiseman commenced by pulling out the patient's loose teeth, then set to work with a series of 'actual' cauteries or hot irons:

> [H]aving his Head held firm, and his lower Lip defended, I passed in a plain Chisel cautery under the *Fungus*, as low as I could, to avoid scorching of the Lip, and thrust it forward towards the Tongue, by which I brought off that *Fungus* and the rotten *Alveoli* at twice or thrice repeating the Cautery; then with Bolt-cauteries dried the *Basis* to a crust. After with a Scoop-cautery I made a thrust at the *Fungus* over-spreading the left Jaw, and made separation of that, and what was rotten of the *Alveoli:* then with Olive and Bolt-cauteries I dried that as well as he would permit.[43]

This patient's surgery was far lengthier and more dangerous than the simple excision with which Wiseman had removed the cranial tumour. As the limits of the patient's 'permission' indicate, it must also have been excruciatingly painful. Wiseman and his contemporaries recorded more of these kinds of operations – lengthy removals including the use of both knife and cautery – than they did simple lumpectomies, despite the fact that these complex procedures were often unsuccessful. The unfortunate Captain, for example, endured several more days of similar treatment, but eventually died when the tumour spread throughout his mouth and into the larynx, an outcome which Wiseman attributed in part to reluctance to allow him 'to keep down the *Fungus* afterwards as it arose' by use of further cautery.[44]

Wiseman seems to have been particularly innovative in his cancer surgeries, and assiduous about recording the most interesting examples. Operations for facial tumours, however, were recorded throughout the early modern period. For example, the 1634 collected *Workes* of Ambroise Paré, which had first appeared in French in 1575, recounted a 'new and never formerly tried, or written of way' by which the author had removed a facial tumour in a 50 year-old man.[45] 'The way is this', instructed Paré:

> The Cancer must be thrust through the lips on both sides, above and below with a needle and threed, that so you may rule and governe the Cancer with your left hand, by the benefit of the threed (least any portion thereof should scape the instrument in cutting) and then with your Sizers in the right hand, you cut it off all at once, yet it must be so done, that some substance of the inner...lippe, which is next to the teeth, may remaine, (if so be that the Cancer be not growne quite through) which may serve as it were for a foundation to generate flesh to fill up the hollownesse againe. Then when it hath bled sufficiently, the sides & brinkes of the wound must be scarified on the right and left sides, within, and without, with somewhat a deepe scarification, that so...we may have the flesh more pliant and tractable to the needle and threed. The residue of the cure must be performed just after the same manner as we use in hare-lips'.[46]

Omitting the hot irons later employed by Wiseman, Paré's operation offered the opportunity to 'rule and governe' this most ungovernable disease. Perhaps tellingly, however, the success of his venture was unrecorded: Paré advanced the method as one by which cancers might be cured without cautery and the associated scarring, but gave no details as to the survival or otherwise of his patient in this case. Despite the uncertain outcome of Wiseman and Paré's procedures, versions of the same were employed throughout the late sixteenth, seventeenth and early eighteenth centuries.[47]

While a number of medical practitioners seem to have been aware of, and occasionally practised, operations for facial tumours, in general cancer surgery reflected the disease's status as paradigmatically afflicting the female breast. Despite its invasiveness, the mastectomy operation was by far the most prominent in medical textbooks, casebooks and advertisements. Most mastectomies followed a similar template: the pulling away of the breast from the body, followed by the removal of the whole breast with a sharp implement. William Beckett's 1711 *New Discoveries Relating to the Cure of Cancers* relates the procedure in brief but excruciating terms:

> Let the Patient be placed in a clear Light, and held steady; then take hold of the Breast with one hand, and pull it to you; and, with the other, nimbly make Incision, and cut it off as close to the Ribs as possible, that no Parts of it remain behind. But if any *cancerous Gland* should remain, be sure to have actual Cauteries of different sizes, ready hot by you, to consume it, and to stop the Bleeding; or otherwise apply, for restraining the Hemorrhage, Dorsels dipp'd in scalding hot *Ol. Terebinth* [turpentine oil] ... then with good Boulstring and Rolling, conveniently place the Patient in Bed, and at night give her an *anodine Draught*, then the second or third Day open it, digest, deterge, incarn and siccatrize.[48]

Beckett's procedure contained several variables which medical practitioners altered according to their own preferences. He provided no instruction, for example, as to what one should use to 'nimbly make Incision'. Most operators favoured a knife or razor, but the Dutch surgeon Paul Barbette noted that some surgeons used needles or hooks and a 'string'.[49] In his 1710 *A Course of Chirurgical Operations*, Dionis suggested one used both, helpfully supplying a diagram of his preferred equipment (Figure 6.3).[50] 'The Chirurgeon', instructed Dionis, 'with Ink traces out the whole Circumference, which is the place where the Incision is to be made':

> [T]hen running the crooked Needle D, across the Body of the Tumour; it is threaded with the String E, whose two ends are tied, and with which he makes a Noose which serves to sustain the Tumour, and in drawing it to separate it from the Ribs...then with Razor F, or a large flat Knife G...the Chirurgeon cuts at the marked Place, and takes off the whole Body of the breast in a short time.[51]

It seems – though Dionis's explanation is unclear – that the string was passed through the base of the breast using the needle (as shown in Figure 6.4, from Scultetus's *The Chyrurgeon's Store-House*). This served to partially separate the breast from the underlying muscle so that it was more stable and could more easily be excised. Kaartinen argues that the

of *Chirurgical Operations.* 247

FIGURE XXVIII. *The* APPARATUS *for the Operation practised on the* CANCER.

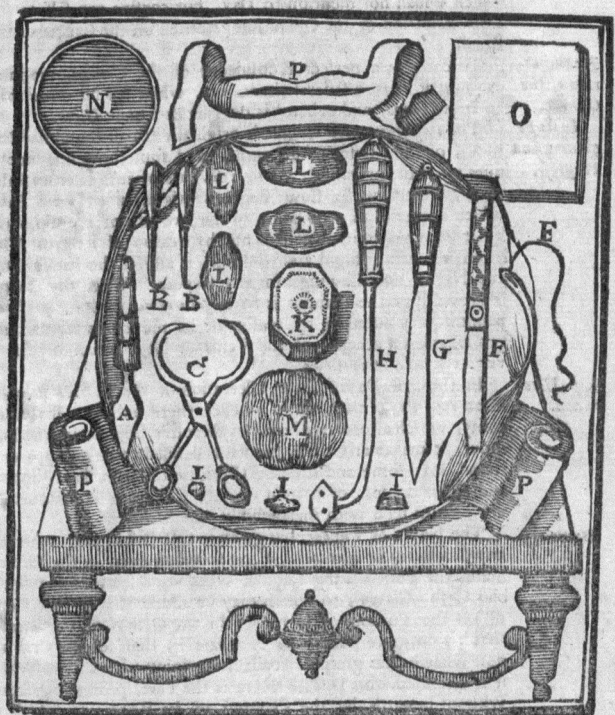

A Cancer is univerfally agreed to be the moft terrible of all *Of the* the Evils which attack Mankind; though Wars and CANCER. Plagues kill in lefs time, they don't yet, to me, feem fo cruel as the Cancer, which as certainly, though more flowly, car- ries thofe afflicted with it to the Grave, withal caufing fuch Pains as make them every day wifh for Death.

'Tis a Difeafe which attacks not only the Breaft, but feveral *Reafon of* other Parts, on which it is not lefs outrageous : It fometimes *its different* R 4 affumes *Names.*

Figure 6.3 Pierre Dionis, *A Course of Chirurgical Operations, Demonstrated in the Royal Garden at Paris* (p. 247), 1710. Copyright of the University of Manchester. This image is open access under a CC-BY NC-SA 4.0 licence.

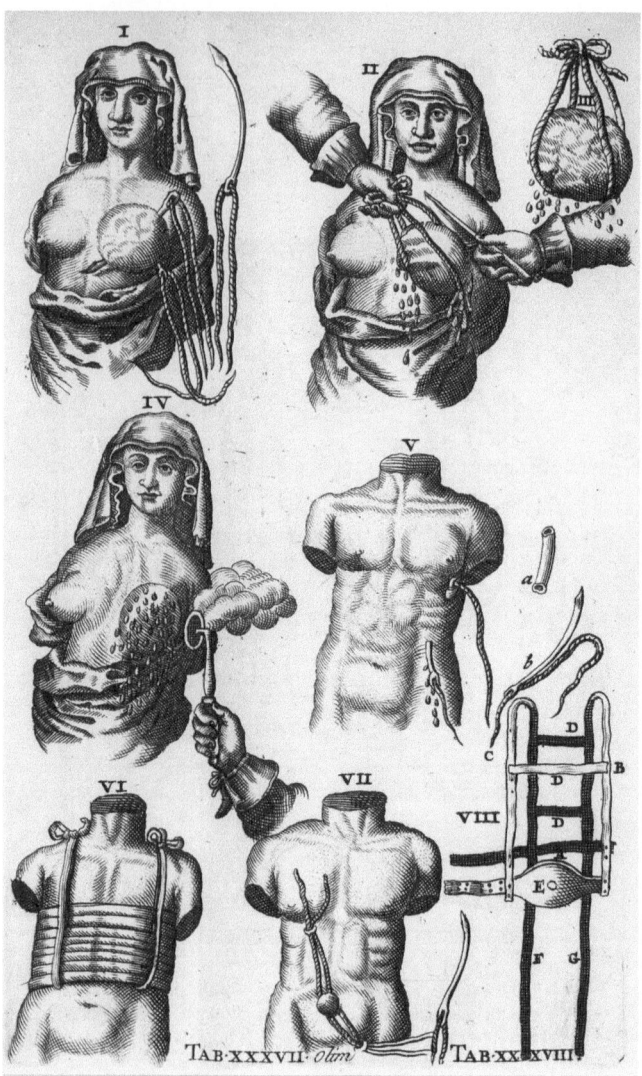

Figure 6.4 Johannes Scultetus, 'Breast cancer operation', from *Het vermeerderde wapenhuis der heel-musters*, 1748. Courtesy of Wellcome Library, London. This image is open access under a CC-BY 4.0 licence.

needle and cord technique was 'in vogue' in the late seventeenth and early eighteenth century, after which it gradually disappeared.[52] In the sources I have examined, however, it seems to have been uncommon.

There were, of course, exceptions to this rule: for example, a surgeon at Saint Bartholomew's hospital, Joseph Binns, took the string method to an extreme. Tying a string around the breast on the morning of 9 August 1648, he 'tied it harder' over the next 13 days until on the 22nd, 'the lower string was through the bigness of a finger, the upper one near to an inch' and he 'with string cut [the whole breast] off in the ligature'.[53] Predictably, however, the patient died a week later: the absence of this procedure from other contemporary texts gives the impression that Binns either misunderstood instructions such as those given by Scultetus, or tried this method as an ill-fated experiment.

In a 'typical' mastectomy, therefore, the surgeon would probably use a knife to cut away the breast tissue. In all likelihood, he would have removed virtually the entire breast down to the chest wall. Dionis described a lumpectomy operation to be used when the cancer was small, palpable and movable, but he was in the minority.[54] Conversely, Beckett recalled observing an operation in which 'a Part of that [pectoral] Muscle was cut away, and the cartilages of Two of the Ribs laid bare, and the patient happen'd to be cur'd'.[55] This too was uncommon, presumably because it increased mortality rates even further.[56] While they were wary of removing too much flesh, surgeons remained mindful of the disease's characteristic malignancy, and repeatedly stressed the importance of removing every trace of the cancer. '[I]t must be all taken away', stressed Bonet:

A Canker once cut doth often come again, 1. When all was not cut out, through timorousness, either in the Operatour, or in the Patient. 2. Because the Arteries that emit this vitious bloud, by reason the less Arteries are cut away from the part affected, must contain more bloud than before, and therefore when they are open, will discharge that bloud upon some other part, whence comes a new Canker. 3. Because there is so much malignity latent in the Body, that a Canker will always grow afresh.[57]

Though the operator could do little about cancer 'latent in the body', he could, it was believed, minimise the risk of recurrence by pressing the bad blood out of the nearby veins and making sure to excise every scrap of cancer either with the knife or cautery. Precisely what means were used to complete the operation and stop the wound from bleeding was mostly a matter of individual choice, sometimes influenced by the constitution and temperament of the patient. Dionis, for example, reported that he had stopped using hot cauteries because they 'make the

Patient tremble' and he could achieve the same result by skilful use of the knife, followed by 'Pledgets' (material pads) and 'astringent powders' to stop the bleeding.[58] In line with contemporary wisdom that closing a wound was dangerous, surgeons generally did not stitch the site of mastectomies or other substantial cancer operations until later in the eighteenth century.[59]

Post-operation, the patient was at high risk of infection, as well as remaining in considerable pain. Occasionally, surgeons would return to treat the wound with hot cauteries again.[60] Whether because this course was intolerable to the patient, however, or because it was ineffective, such extended treatment was fairly uncommon.[61] Instead, surgical texts often recorded either the authors or their colleagues administering prescriptions with soothing and anti-inflammatory properties, as well as some potent analgesics. Wiseman, for example, prescribed one mastectomy patient a 'Pearl-Julep' 'to refresh her fainting spirits', and the next day she was given 'distilled milk', containing, among other ingredients, gentian, rose, agrimony, cinnamon and veronica.[62] In 'extremity of pain', he recorded, she was to be given a drink made with theriac, a concoction which usually contained opium and snake venom.[63] In many cases, it appears that surgeons monitored their patients closely in the days after surgery, and remained aware of the potential for infection or a recurrence of the cancer for months, even years. For their part, patients were advised to be constantly on the lookout for new tumours, and told they 'must not discontinue the use of internal Remedies for some Years, lest a Fresh tumour should break out in some other Part, and produce a new Cancer'.[64]

Descriptions of early modern cancer surgery showed a relative homogeneity, pointing to the existence of established operative conventions, and to a steady stream of patients who were willing to put those conventions to the test. Despite their exceptional invasiveness, such operations were broadly intuitive, aiming for a golden mean between extirpating the cancer thoroughly and minimising dangerous blood loss. Interestingly, they were also united in the way in which they described the process of operation. Surgeons, as we have seen, vividly portrayed the sufferings of their patients prior to surgery. They also, to a lesser extent, showed empathy with the pain and shock experienced by patients after a major cancer operation. Descriptions of the operation taking place, however, showed no such personal attention. Rather, they were characterised by an anatomical emphasis in which the person under the knife was consistently reduced to the sum of his or her parts. The reasons for, and effects of, this phenomenon are the subject of the remainder of this chapter.

6.3 'What then can we think of this shameful Undertaker [?]'[65] Competing narratives of cancer surgery

Reading early modern instructions for and accounts of cancer surgery is a stark reminder of just how dangerous and painful these operations must have been. Clearly, some patients summoned the strength to undergo such procedures because they believed surgery was the only option left to relieve their sufferings, and prevent their premature deaths. As we have seen in the previous chapter, however, the recourse of desperate patients to such extreme measures may in fact be less remarkable than the willingness of medical practitioners to administer them. For surgeons, as for physicians, undertaking invasive and bloody procedures was a course often fraught with doubt and difficulty. Many surgical texts show that operators were traumatised by the screams and struggles of patients in agony under the knife. Moreover, when they attempted anything but the most superficial excisions, surgeons risked killing or maiming the patient, incurring serious and lasting damage to their reputations and hence their livelihoods.

In these grim circumstances, several of the factors which motivated early modern surgeons to conduct cancer operations were clearly linked with those which compelled sufferers to consent to this course. Firstly, operators were all too aware of patients' often chronic and unremitting pains. Cancer sufferers' pleas for relief at any cost clearly rang loudly in the ears of many medical practitioners. Secondly, cancer in some senses 'invited' surgical intervention by dint of its seemingly evil and rebellious nature. To the early modern mind, cancer was hostile and malign: an alien to the body repeatedly imagined as deliberately resistant to cure, and aligned with evil influences in the world at large. For medical practitioners as well as patients, surgery offered a chance to reach into the body and remove the interloper, and the language of surgical textbooks often represented (and reinforced) an adversarial relationship between medical practitioners and cancer. In his 1583 *The Method of Physick*, for example, Philip Barrough counselled medical practitioners to 'devide the good from the evill' when excising cancers.[66] A text by Jacques Guillemeau and 'A.H.' similarly advised that the 'reliques' of the disease be 'abolish[ed]' – language that must have echoed particularly loudly in post-Reformation English ears.[67] Repeated injunctions to remove all the cancer not only advised on clinical practice, but reflected and reinforced appealingly tangible and symmetrical ideas of cure: that the body could be restored by cutting into it, and the disease of burned humours could be quelled with burning iron.

Surgeons thus responded to both the physical reality and the rhetorical construction of cancer as a fearsome, evil disease. Furthermore, in many surgical texts, it is clear that discussions about cancer operations constructed those surgeries as not only compassionate, but contributing to medical knowledge and the 'progress' of surgery more broadly. In the adversarial drama played out between surgeons and the cancers they sought to eliminate, there was a distinct sense of intra-professional (and largely homosocial) cooperation as well as competition. This was partly a matter of necessity – surgeons needed assistance to keep the patient held steady, pass instruments, heat iron cauteries and apply 'pledgets' or pads to stem bleeding. To a greater extent than physic, surgery was a trade learned through apprenticeship, and many operators could have expected to have one or more such charges in attendance.[68] In a broader sense, surgeons were 'apprenticed' to the ancient and medieval medical writers whose advice they often cited. Demaitre notes the influence of Rhazes (Muhammad ibn Zakariyā Rāzī, 865–925 AD) and Galen on medieval discussions of cancer surgery by Avicenna and Lanfranco, who were in turn frequently cited by seventeenth-century writers.[69] Surgeons undertaking such operations could therefore feel that they were contributing in their turn to a patrilineal development of knowledge.

Even when they were not required for practical purposes, it is clear that many experienced surgeons and other medical practitioners attended and assisted at cancer surgeries, particularly mastectomies and invasive facial operations, out of professional curiosity or camaraderie. Wiseman, for example, recorded that he had examined and operated on cancers in conjunction with, or in the presence of, other medical practitioners including Walter Needham, 'Mr. Nurse', Doctor Bate, Doctor Thomas Cox, Doctor Micklethwaite, Jacques Wiseman (his 'kinsman'), and Mr. Hollier, Mr. Arris, Edward Molin, Mr. Troutbeck and Mr. Shunbub (all chirurgeons).[70] Likewise, at the mastectomy observed by Reverend John Ward, which took place over several days, two surgeons, 'Clerk, of Bridgnorth' and 'Leach, of Sturbridg[e]' operated, while Walter Needham arrived too late on the first day, but 'staid...to see it opened' again the next day, and 'Dr. Edwards' marked with ink 'the way how and where it should be cut'.[71] That surgeons were seemingly so keen to be involved with cancer surgeries, despite the risks to their reputations in the event of a patient's death, shows how fascinated they were by these procedures. Their attendance at and detailed recording of operations with a novel pathological or methodological element also suggests that they saw cancer operations as potentially perfectible: a coup which, if achieved, would undoubtedly bolster the claims of many surgeons that their craft should be considered a noble profession, equal to that practised by university-educated physicians.

Surgeons who dwelt on the technical improvement of cancer surgeries clearly believed that in the long term, operative advancements could benefit both practitioners and patients. For the individual sufferer, however, this 'long view' could reach unsettling extremes, allowing surgeons to ignore the suffering of individual patients in the service of curiosity, learning or fame. Notably, in scholarship on early modern dissection and vivisection, Sawday, French and Egmond have all noted an imaginative connection between these occupations and that of the surgeon. Concomitant with the intense interest in dissection and anatomy during the early modern period, they argue, was a suspicion that living humans might be next under the curious anatomist's hand.[72] For instance, citing Edward Ravencroft's *The Anatomist* (1697) and Thomas Nashe's *The Unfortunate Traveller* (1594) as examples, Sawday contends that the idea of a *living* anatomy possessed a peculiarly compelling horror for early modern dramatists, and that '[i]magining one's own dissection was a device unique to early-modern culture'.[73] It is by no means certain that this fear was unfounded. Egmond mentions 'some evidence of vivisection on human beings', while French notes that '[r]umour...had it that at least two Renaissance anatomists succumbed to temptation and ventured into human vivisection'.[74] As Richard Sugg observes: 'Available data indicates that almost no one was prepared to advocate human vivisection during the Renaissance. By contrast, however...various figures seemed ready to believe that the practice might be carried out by their contemporaries'.[75] Moreover, it was seemingly accepted that if anyone was to venture into vivisection, it would be surgeons, rather than physicians. First published in 1605, Michael Drayton's 'Sonnet 50' vividly imagined that 'in some countries, far remote from hence', condemned criminals might be used as experimental subjects by surgeons, who would

First make incision on each mastering vein

Then staunch the bleeding, then transpierce the corse,

And with their balms recure the wounds again

Then poison, and with physic him restore

Not that they fear the hopeless man to kill

But their experience to increase the more. (l. 6–11)[76]

As Sugg observes, Drayton's fears might have been founded, in part, upon his observation of surgeons' 'necessary, temporary detachment from human suffering', a trait which 'threatened to harden into a permanent and dominant identity in the perception of the lay public'.[77]

Even if they were not explicitly associated with anatomists, surgeons undertaking invasive operations were bound to find their narratives of progress interrupted by the uncomfortable fact of patients' suffering under the knife. The problematic nature of the surgeon's craft, which both healed and hurt, has been noted by several historians of early modern and medieval medicine. Andrew Wear's 'Medical Ethics in Early Modern England', for instance, describes the difficulty of drawing a line between treatments which harmed and those which helped patients, while in her reading of Henri de Mondeville's medieval surgical works, Pouchelle notes that Mondeville himself admitted that 'surgeons have a reputation for cruelty' and 'the surgeon who refuses to be considered as an executioner or public tormentor would become a laughing-stock among "ordinary uneducated people"'.[78] One common concern among surgeons was that they might be perceived as over-eager to employ the knife, and hence, as one ship's-surgeon cautioned, be 'esteemed Butcher-like and hateful'.[79] Cancer surgeons were, it seems, particularly vulnerable to accusations of cruel, callous or incompetent conduct which allied them with the anatomist, torturer or butcher. The operations they carried out were some of the most lengthy and dangerous undertaken during the early modern period, particularly in the case of mastectomy. Furthermore, these operations were not always immediately and visibly necessary. It was easier to decry a surgeon removing a superficially healthy breast which contained palpable tumours than it was to quibble with an operator who caused similar pain while removing a bullet or amputating a mangled limb.

In this suspicious climate, the language with which some surgeons chose to describe their operations suggests that they, too, were uncomfortable with the pain they inflicted. In some cases, it is clear that cancer operators preferred, or perhaps needed, to view the person under the knife as a specimen rather than a thinking, feeling patient. Many accounts of surgery show operators focussed on their relationship with other practitioners or with the 'rebellious' cancer, to the exclusion of the patient as subject. Wiseman's description of a mastectomy performed on a 'Country-maid', for example, contains no details about the patient other than her occupation, age and the initial appearance of her breast.[80] It does, however, give a detailed account of 'the experimenting of the Royal Stiptick liquor' (designed to stop bleeding), the arrival and involvement of Needham and Jacques Wiseman, and Richard Wiseman's attendance on some 'friends' who wished to see the new stiptick.[81] From the time the operation is resolved upon, to when it is completed, the whole body of the patient is never referred to, but is only manifest through the breast, the tumour and the blood issuing out. This erasure of the patient was by no means confined to Wiseman. Looking again at Figure 6.4, for example, one sees in Scultetus's diagram the depersonalization of the woman under the knife. In the top left-hand image, we view the patient, looking oddly serene as the needle is passed through her breast, her hair covered and seemingly armless. The accompanying text describes 'a *Breast* affected with an ulcerated Canker', effacing the subject attached to that breast.[82] In the next picture, the hands of the surgeon[s] descend as if from the heavens to remove the breast, and in the third, the (literal) dissociation of patient from cancer is complete as the amputated breast hangs, detached, 'weighing six physical pounds'.[83] The pictures marked V, VI and VII on the same page are meant, according to the text, to represent treatment for a fistula, bandaging of the thorax and correction of a hernia.[84] Their continuous numbering with the mastectomy pictures, however, rather implies a continued improvement – that the ideal or corrected body is one in which both subjectivity (the face) and femininity (both the breasts) are absent.

The uneasy relationship between femininity and cancer surgery is discussed later in this chapter. In relation to surgeons' self-construction as compassionate and progressive, however, it is evident that taking patient subjectivity out of the equation in texts on cancer surgery served several purposes. First, while surgeons acknowledged the pain of surgery when discussing the decision to operate and the proper provision of aftercare, excluding the patient at the moment of greatest suffering – under the knife – made it easier for surgeons to construct themselves and their activities in their own, flattering, language, rather than the fearful or suspicious terms in which they were often criticised. Furthermore, the exclusion of a patient's thoughts, feelings and personality from textual representations of surgery mimicked the detachment which was deemed necessary in order for surgeons to do their job. In her work on medical dispassion in early modern England, Payne describes at length the trauma and difficulty inherent in operating upon conscious patients.[85] Lengthy cancer operations were particularly distressing for all involved; as one *Medical Dictionary* advised, women undergoing mastectomy might 'shriek and cry in a manner so terrible, as is sufficient to shock and confuse the most intrepid surgeon, and disconcert him in his operation.'[86] Under such circumstances, the surgeon had to 'equip himself in all the steps of his operation, in such a manner, as if he was deaf to the moving groans, and piercing shrieks, of the tortur'd patient.'[87] In fact, as the *Dictionary* implied, the best sort of patient would be a silent, unfeeling carcase, such as young surgeons sometimes practised upon.[88] Confirming this fantasy, and relaying instructions for mastectomy, Dionis informed young surgeons that '[t]his Operation is easier than is imagined before 'tis performed; for the Breast separates as easily from the Ribs, as when we divide the Shoulder from a Quarter of Lamb'.[89] His statement, seemingly meant to reassure, tacitly acknowledged the dread with which some operators must have approached this procedure, and the mental tactics employed to overcome it.

Representations of cancer surgery thus consistently engaged with the potential of that operation both to help and harm. Where cancer surgeons might try to efface the dangerous and painful nature of their interventions, however, other medical practitioners had no such qualms. For every author who provided accounts of or instructions for cancer operations, there were many more writers – often physicians, but sometimes lay onlookers or surgeons writing against their perceived inferiors – who accused cancer surgeons of conduct which was at best careless and at worst positively evil. In a 1703 publication from 'T.D.' on the 'Abuses' committed under the name of chirurgery, for example, the author singled out one surgeon's cancer operations for particular attention.[90] This operator was, it seems, moderately famous for mastectomy operations in particular: T.D. stated that 'I make no question but you have hear'd of one who calls himself the un-born Dr'.[91] The doctor's practice, wrote T.D., was 'monstrous': 'The Number of Womens Breasts, which this man has cut off within these few Years is scarce to be believ'd: And yet ... he cannot produce One, where there was a true ulcerated Cancer, that is now living to tell Tales of Him'.[92] Given that cancer was widely acknowledged to be difficult if not impossible to cure, 'what then can we think', asked the author,

> of this shameful Undertaker, who makes no more of taking off a Breast (altho' no otherwise than a Butcher might do the same) than some Persons do to pair [pare] their Nails, so that scarce any thing of a distemper'd Breast is presented, but the poor Woman is frighten'd out of her Wits, with the dismal Sentence pronounc'd of its being Cancerous.[93]

For T.D., the activities of the 'unborn Dr.' could not be viewed as compassionate or progressive. Instead, the casting of the surgeon as 'Undertaker' in this account explicitly opposed the operator's self-construction as a preserver of life. Moreover, naming the doctor as a 'Butcher' who cut up women as readily as he cut his nails subverted surgeons' emphasis on the professionalism of their craft and prefigured, in distorted form, Dionis's assertion that mastectomy might be as easy as dividing up a shoulder of lamb. Undoubtedly, there were foolish or unscrupulous practitioners to be found in every kind of surgery. However, T.D. implied that cancer surgery was an area in which unscrupulous practitioners could make their mark particularly easily, because women were so afraid of the disease that they could easily be manipulated into undergoing unnecessary operations. As someone who apparently grew his own coffers by doing physical harm to his patients, this 'Dr.' might even be viewed as malignant in his own right.

These accusations were damning and imaginatively compelling ones, calculated to strike a chord with contemporary fears about the motivation and competency of surgeons. Even 'T.D.' did not argue that surgeons actually enjoyed inflicting pain. However, the obvious agony of the cancer operation, combined with surgeons' reluctance to acknowledge that pain in their medical writings, inevitably led to accusations that those who carried out these procedures were more interested in personal gain and professional advancement than in the humanity of their endeavours. As a profession, surgery could not escape the fact that only the intent to heal definitively separated the surgeon from the torturer, and only a successful result distinguished him from the anatomist. That cancer surgery came in for particular scrutiny in this regard was a product of several factors. These operations were, as I have shown, unique in their invasiveness and the fact that they were undertaken at the patient's behest or with their pre-obtained consent. Furthermore, belief in the evil, quasi-ontological nature of cancer fostered the desire to extract this interloper from the body in a way unmatched for other diseases. Even contemporary surgeons identified cancer as a disease particularly likely to provoke dangerous 'experiment' with 'bold and rash' pharmaceutical and surgical methods, precisely because it was such a mysterious and fascinating malady to medical practitioners.[94] Throughout the early modern period, it seems, both surgeons and those who observed their activities knew that therapeutic encounters with cancer and the preservation of humanity – in both patient and operator – could not easily coexist.

6.4   'And in such searching wounds the surgeon is / As we, when we embrace, or touch, or kiss': cancer surgery and gender relations[95]

All kinds of cancer operation were controversial. The dangerous and invasive nature of such procedures led to much criticism of those who dared to undertake them – mostly from other surgeons and medical practitioners convinced of the futility of such interventions. Occasionally, however, those surgeons who carried out cancer operations tacitly revealed their own anxieties about opening up the body. These anxieties related, to a striking degree, to female patients, and mastectomy operations. Moreover, they cut both ways, involving the possible subjugation of female patients and emasculation of male operators.

Early modern medicine in general was often imagined as a sexually charged pursuit. The fact that male medical practitioners possessed intimate knowledge of the female body made their craft, as Roy Porter observes, 'inescapably associated in the public imagination with carnal knowledge'.[96] Erotic prints and poems, he notes, commonly 'exploited "medicine" as a double entendre, cover, or euphemism for sexual opportunism'.[97] Physicians and apothecaries, however, were generally employed in diagnosing complaints and prescribing medicines rather than physically manipulating their patients. It seems evident that surgery, which was necessarily a tactile and intimate encounter, should be even more vulnerable to accusations of sexual misconduct, and tensions ran particularly high when (usually male) surgeons operated on female patients. As a paying customer, any patient, male or female, possessed a high degree of agency over their treatment. Kaartinen has shown that for cancer in particular, many women had substantial knowledge of the surgical and medical treatments available to them, and readily asserted their own opinions as to their treatment.[98] Conversely, however, Laura Gowing notes that simply being touched could undermine an early modern woman's social status.[99] When exposed to touch in inappropriate ways – touched by too many people, or the wrong sorts of people – women's bodies risked being deemed 'common', and compared to the ultimate 'common' body, that of the prostitute.[100] Male surgeons touching female patients (and likewise, patients being touched) were, therefore, precariously positioned. Surgeons exercised a peculiarly acute power of touch capable of inflicting not only social but mortal physical damage. At the same time, their access to the body was, as I shall demonstrate, contingent and uncertain.

As described in the first section of this chapter, many cancer patients chose, even demanded surgery, in full knowledge of the likely pain and danger to their life. Some surgeons consented only reluctantly in view of the traumatic nature of the procedure and the attendant danger to their reputations. However, this was not always the case. Several accounts from medical casebooks and instructional texts recall situations in which surgeons tried, unsuccessfully, to persuade patients to undergo surgery. These situations related almost exclusively to women, and were frequently framed in gendered or sexualised terms. In 1698, for example, *The Compleat Midwife's Practice* recounted the story of an unnamed woman with breast cancer, which was becoming gradually worse.[101] '[A] skilful Surgeon', recalled the authors,

> refused to open it, but advised the best he could to give her ease, and promised to come to her, if after it brake she would send for him. Some Months after she sent for him, and shew'd him a great quantity of curdled matter newly burst forth; the Breast was lank, but very hard *Glands* lay within, and ... there were some *tubercles* that required to be eradicated; to which purpose, he design'd to have slit open the *abscess*, and to have pull'd away the Cancerated *Glands*, but she would not permit him so much as to enlarge the orifice; upon which consideration he left her, and she died within half a year after.[102]

The authors' sympathies clearly lay with the 'skilful' surgeon in this bizarre tale. As well as an exhortation to readers to submit to the advice of their surgeon, however, the account reads as a gendered power struggle centred upon the surgeon's thwarted desire to penetrate the unnamed 'orifice'. Stressing the anatomical terms in the story – *'tubercles'*, *'Glands'* and *'abscess'* – the author tries to emphasise clinical details of the case, but his narrative, like the unnamed surgeon's plan, is continually disrupted by a female who gives her opinions clout by denying access to her body. In certain lights, a woman's reluctance to have her breasts examined or treated by a male practitioner could be construed positively, as an instance of proper feminine modesty. This was, for instance, the case for the writer Mary Astell, whose reluctance to seek treatment for her cancer was represented in a posthumous biography as exemplifying her patience and fortitude.[103] However, in late sixteenth-, seventeenth- and early-eighteenth-century texts, reluctance to undergo surgery which had been recommended by a medical practitioner was more likely to be depicted as an example of womanly foolishness and obstinacy. Despite the power they wielded during an operation, surgeons were service providers, and were not, in principle at least, allowed to coerce or bully their customers into a procedure. Their opinions were automatically overruled by those of their customer, the reluctant patient, and this clearly sat uncomfortably with some surgeons in a society which traditionally privileged the voices and judgements of men.

The refusal of 'permission' by the female patient in *The Compleat Midwife's* account was elsewhere formulated as a failure to 'submit', a term which was used in texts on cancer exclusively to describe women who were uncooperative with their medical practitioners.[104] For instance, Daniel Turner recalled in 1714 that encountering a patient with facial cancer, 'I told her if she would submit to the hot Iron, I would serve her so far as I was able, believing that the most likely Remedy for so obstinate a Disease'.[105] The patient was, understandably, frightened by the prospect of the 'fiery Tryal' and refused Turner's intervention in favour of remedies from an 'Empirick'; predictably, it was reported that the cancer had now spread over her face.[106] Once again, the encounter was framed in loosely sexual terms, as to 'serve' a woman could also mean to act as her lover or impregnate her.[107] This aspect of the surgeon-patient relationship was even more prominent in an account by Dionis of the treatment of Madam de Montreuil, a lady who sought his advice whilst he was travelling around France with some colleagues.[108] This lady, unlike Turner's patient, was easily persuaded that surgery was necessary for her breast cancer. However, circumstances meant that Dionis was unable to operate. He recorded: 'She would have desir'd me to have perform'd the Operation; but that she had then her Terms, and having no more than two days to stay at *Marseilles*, I could not satisfie her'.[109] It was not unusual to delay an operation until after a patient's menses. However, the language of 'desire' and 'satisfaction' here connected surgical and sexual performance, particularly as sex during menstruation was commonly believed to be unhealthy.

In scenarios like these, the access of a male surgeon to a female patient's body was implicitly framed in sexual terms. The narratives presented by medical practitioners depicted any resistance to their desires, therapeutic or otherwise, as foolish misjudgements – perhaps characteristic of ignorant and fearful women – which ended badly for the intractable patient. It should be noted that there was no suggestion in early modern texts, medical or non-medical, that surgeons actually experienced sexual gratification from operating on women's breasts. Nonetheless, violence, sexual gratification and surgery were somehow allied, and mastectomy – a dangerous, body-altering operation – was naturally susceptible to such associations. For example, when painting Saint Agatha's tortures, numerous sixteenth- and seventeenth-century artists depicted her tormentors using the surgical instruments of the period.[110] Perhaps this is unsurprising: after all, questions of power and violence attached to mastectomy have long been a focus of modern cancer studies such as the tellingly titled *A Darker Ribbon* and *The Breast Cancer Wars*.[111] Examining a nineteenth-century image of mastectomy, Bridget L. Goodbody makes a similar link between different forms of power over the female body. In *The Agnew Clinic* (Figure 6.5), she argues, one can trace an 'erotics of sadism', in which the 'supine and helpless position' of the patient 'creates the sense of her willing submission [to the doctors]...even to the point of willingly placing herself in a violent circumstance from which she cannot escape'.[112] Crucially, the semiotics of the situation are not diminished by the operators' good intentions:

> [T]he surgeons knew that the patient's fragile life rested very precariously and tenuously in their hands. Taken to the extreme, this thought prompts the question: How far could they rationally and almost ritualistically violate her body to establish their power over her and her cancer without killing her? Such questioning is not intended to imply that the doctors derived pleasure from her pain.[113]

As such analyses highlight, where a gathering of men takes place over a female body, questions of 'violation' may arise even where it is clear that the surgeons involved did not purposely exploit that body or gain pleasure from the scenario. Moreover, in a heteronormative society, the

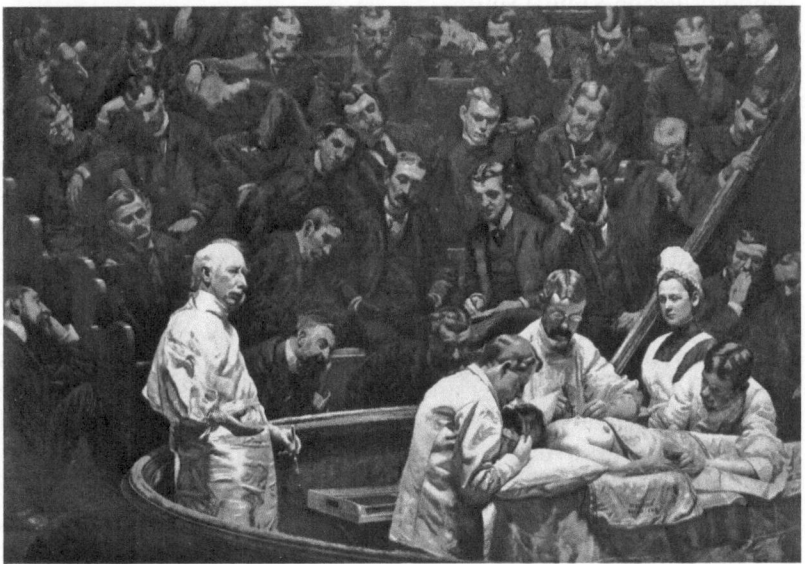

Figure 6.5 The Agnew Clinic. Artist/maker unknown, After Thomas Eakins. Photogravure, c. 1889. Image: 7 7/8 × 11 in. (20.0 × 27.9 cm). Courtesy of Philadelphia Museum of Art: Gift of Samuel B. Sturgis, 1973. This image is open access under a CC-BY 3.0 licence.

'dynamics of inequality' created by such a scenario were readily sexualised.[114] The very fact of a female patient placing her life quite literally in the hands of a person of the opposite sex carried an erotic charge in a culture in which – as was true of early modern English society – submission and subordination were indexed to good 'femaleness'.

The peculiarly intimate access to the female body and breasts afforded by cancer surgery might thus be read as connoting sexual desire or domination even though it was never suggested that operators actually viewed their work in this light. Tales of women who refused to comply with surgeons' advice were more common than the equivalent for men both because females made up the bulk of cancer diagnoses and thus surgical cases, and because their assertion of bodily agency was particularly significant in a broadly patriarchal society. This is not to say, however, that cancer surgeries on women were experienced as unproblematic exercises of male power. Cancer was, as we have seen, a disease known for its malignancy, secrecy and resistance to cure. In surgical encounters with the female body, these characteristics could play out in ways that highlighted issues of gender and power, and this was emphatically the case in one unusual but instructive tale, that of London surgeon Samuel Smith.

Cited at length in Chapter 4, Samuel Smith's story epitomises the double danger posed to male surgeons from involvement with the 'cruel' malignancy of cancer, and the troublingly illimitable female body. '[A]t the cutting off of a large Cancerated Breast', it was reported, Smith, a surgeon at St. Thomas Hospital in Southwark, 'had (after the Breast was off) a Curiosity to taste the Juice, or Matter contain'd ... which he did by touching it with one of his Fingers, and then tasting it from the same with his Tongue'.[115] Tasting a patient's bodily fluids was not unknown in early modern diagnostics, and F. David Hoeniger notes that 'sour and sharp' tastes in blood were thought to indicate an excess of melancholy humours therein, consistent with the outcome in this case.[116] Nonetheless, tasting amputated tissue was unusual, and the fact that the 'large' breast belonged to a patient who may have been conscious under the surgeon's hands once more highlights the uncomfortable proximity between medical and sexual touching.

The most dramatic part of this story, however, was still to come. Immediately upon tasting the breast, the surgeon complained that the matter had a persistent and permeating acrid taste. Within 'a few months' the surgeon found himself in 'a Consumption, or wasting pining Condition' and died soon afterward.[117] Smith's misfortune was taken by the anonymous author as an indication of the quasi-poisonous malignancy of cancers. However, it was also clearly gendered. The cancerous matter, for instance, 'pierce[d] through the whole substance of his Tongue, and passed down his Throat', rendering this 'very strong Man' as weak as the woman upon whom he had operated. Moreover, the author's emphasis on this transformation pointed to the corrupting potential of the similarly illimitable female body. As Paster has argued at length in *The Body Embarrassed*, the female body was thought to be characterised by superfluity, leaking and disorder, expressed through the involuntary and incontinent shedding of bodily fluids including tears, milk, urine and blood.[118] Smith's plight, which rendered him 'wasting' and 'pining', realised the possible dangers of coming into contact with female excreta, compounded by the noxious and malignant substance of the cancer.

While Smith's subsequent illness was understood to result from his ingestion of the cancerous 'juice', the story also gestured to less obvious kinds of contamination by the female body. In *The Body Emblazoned*, Sawday notes that anatomists risked emasculation as they opened up women's bodies. 'Once the body has been partitioned and its interior dimensions laid open to scrutiny', he writes, 'the very categories "male" and "female" become fluid, even interchangeable'.[119] This concern accorded with broader discourses of the period which were concerned with infection and contagion, including through the air or by sight.[120] Writing on 'contagious sympathy' in Shakespeare, for instance, Eric Langley notes the mingling of science and rhetoric which fostered belief in infection by sight, 'a material thread of connection or contagion between viewer and viewed'.[121] Barbara M. Benedict similarly identifies curiosity – the trait which caused so much trouble for Smith – as 'a perceptible violation of species and categories', which might include violation of proper gender attributes.[122] Once again, these concerns were emphasised by cancer's well-known tendency to spread and resist medical intervention, as well as remaining 'hidden' prior to ulceration. Like cancers, women's bodies might be viewed as hazardous when they remained 'secret', and even more dangerous when opened up to the medical practitioner's view.

Cancer surgery, and mastectomy in particular, was difficult and dangerous. In such circumstances, it is easy to see how female patients might be dominated, even inadvertently, by male surgeons. These stories, however, demonstrate that the gender relations attendant on cancer surgery were often more complex than one might expect. As we have seen, male surgeons carefully constructed their craft as compassionate, progressive and professional. Real-life women, with their garrulous voices and unbounded, unfathomable bodies, threatened to bring that edifice tumbling down.

Conclusion

Cancer surgeries were undoubtedly perilous operations, potentially lethal for the patient and professionally damaging for the surgeon. In addition, they were clearly intensely traumatic procedures, causing almost unimaginable pain of which medical practitioners were uncomfortably aware. The fact that such surgeries were nonetheless undertaken throughout the early modern period serves as testament to the agony and debility generated by growing tumours or ulcers. Looking at the language in which surgeons described cancer operations also reveals how far they imagined these procedures as part of a new, professionalised kind of surgery, in which collaboration and competition fostered improvement and innovation. Cancer surgeries served as a focus for these narratives for several reasons. There was a steady demand for tumour removals and mastectomies, such that a relatively standardised method could be established, a common ground for medical discussion. Cancer surgeries were, in a loose sense, elective surgeries, not undertaken on an emergency basis. This meant that surgeons could more readily go to view or participate in complex operations, and patients entrusted surgeons with their lives in an explicit and premeditated sense. Perhaps most significantly, the 'nature' of cancer – its status as malign, rebellious and alien to the body – encouraged an adversarial approach to the disease in which surgery offered the alluring prospect of extirpating the intruder.

These factors combined to ensure that cancer surgeries continued, and steadily increased, throughout the eighteenth century and beyond. Behind these larger narratives, however, individual patients and practitioners experienced surgery in ways that were terrifying, confusing and sometimes frustrating. One of the most curious aspects of early modern cancer surgery is the fact that not a single text I have examined mentions the change in bodily appearance effected by mastectomy, even obscurely. For those who survived this perilous operation, it seems that surgeons were reluctant to confront the possible costs of their success, or to undo the detachment from their patients which allowed them to carry out, and construct as progressive, such risky procedures. Of fables of Amazonian mastectomy in the early modern period, Paster speculates that

> Mastectomy ... implies the Amazon's crucial bodily heresy at least by comparison with the many claims, material and symbolic, on womb and breast in early modern culture – the heresy visibly to control their own bodies, to regulate their own reproductivity, and to offer

a model of self-government in which reproduction and nurture are only two of several forms of service and productive activity.[123]

For the early modern woman, whose mastectomy was a forced choice, one-breasted existence was unlikely to represent a rejection of contemporary gender roles. Nonetheless, her altered body perhaps signalled to others the courage with which she had decided to assert control over her diseased body – even if that agency came at a high price.

Conclusion: 'Death Is Only Their Desire'

This book began with the gruesome record made by Reverend John Ward of a mastectomy operation carried out on 'Mrs Townsend'. In 1666, Ward added the following account:

> Mrs. Townsend, of Alverston, being dead of a cancer, Mr. Eedes and I opened her breast in the outward part, and found it very cancrous; it had been broken, and a mellicerous part was yet remaining when we saw it, which being launct, yielded two porringers full of a very yellow substance...The flesh that was growne againe, after part was taken out, was of a hard gristly substance, which seemed very strange. The ribbs were not putrefied as we could discerne, nor anything within the breast of a cancrous nature, for we runne the knife within-inside the breast through the intercostal muscles. Dr. Needham hath affirmed that a cancer is as much within as without the breast, and he hath seen a string, as I was told, going from the breast to the uterus. I suppose it was the mammillarie veins full of knotts which were cancrous, and hung much like ropes of onions. The cancer was a strange one, as was evident; we wanted spunges and other things convenient, or else we had opened the cavitie of the breast.[1]

Despite (and sometimes because of) the best efforts of surgeons, physicians, apothecaries and empirics, most cases of cancer in the early modern period would, like this one, end in death. In many cases, therefore, people diagnosed with cancer chose to avoid the rigmarole and discomfort of special diets, medicines and caustic salves, or the pain of operations like the one Mrs Townsend endured, and instead follow a palliative course in which they aimed only to delay death and make their illness and demise as painless as possible. Ward made no record of the measures which might have been taken to help Mrs Townsend achieve such a 'good death' after all her sufferings, but we can guess at what they may have entailed. Palliative cures were typically based upon cooling, analgesic remedies for consumption or topical application, often containing ingredients such as plantain, nightshade, scabious and rose.[2] For the later stages of cancerous disease, many medical practitioners admitted that they prescribed increasing quantities of opiates such as laudanum, which despite their addictive properties could offer 'very great comfort' to patients in the last stages of disease.[3] Palliative care did not attract the same level of attention as was given to descriptions of, and 'cures' for, cancer. Moreover, it was not usually specific to cancer. Given the number of morbid diseases to which one might fall victim during the sixteenth, seventeenth and eighteenth centuries, some variety of pain relief was a basic element of medical practice, and could be found described in texts on everything from pox to gout.[4] Nonetheless, it seems likely that outside the remit of medical writings, many patients would have eschewed the radical 'cures' described by surgeons and physicians in favour of a comfortable existence with the chance 'not to dye the sooner, because of that Cancer'.[5]

Moreover, like surgical and pharmaceutical 'cures', end-of-life care for cancer was not divorced from cultural and imaginative constructions of the disease. Ambroise Paré recorded that he had decided upon a palliative cure for one patient 'fearing to irritate this Hydra, and cause it to burst in fury from its lair'.[6] His fear clearly had much to do with the construction of cancer as a purposely malign 'alien' to the body. Likewise, when comparing cancer with the new craze of duelling among the aristocracy, one polemic writer drew on the notorious intractability of the disease to explain that

> as the case stands, the best way with it, is to treat it like a wild and inverterate Cancer … to let it alone, and use no other means, than that of keeping it clean, and making it as easy as we can, since tampering with it can do no good, but in all likelihood only enrage it, and give it an occasion, by showing its Strength, and the Undertaker's Weakness, to encrease its ill Effects, and spread the more and faster.[7]

It seems that cancer was a disease for which palliative treatment was often acknowledged as the only sensible option, given the disease's continuing ability to expose 'weakness' in the practice of even the most eminent medical practitioners. Indeed, this opinion was reiterated by numerous medical practitioners even as they supplied details of the miraculous cures they had effected using surgery and pharmaceuticals. As I noted in my Introduction, it is clear that medical texts did not always reflect everyday practice. Moreover, in common with many aspects of the construction and experience of cancerous disease, the voices of sufferers are almost entirely absent from written accounts, and they disappear from view after attempts at cure have been abandoned. Intriguingly, Gideon Harvey observed in his writings on venereal disease that in one terminal case '[the sufferer's] dearest Friends out of Commiseration perswaded him rather to chuse Death by some Poison, to determine his misery'.[8] It is impossible to tell how many cancer sufferers, being prescribed increasing quantities of opiates, might have chosen to similarly 'determine' their fates.[9]

Mrs Townsend's post-mortem thus provides an appropriate conclusion to this book. During her mastectomy operation, her status as an object of fascination coincided uncomfortably with her subjectivity, the remarkable way in which she 'endured soe much' under the knife and elicited the horrified, fascinated admiration of those who witnessed her pains. In this second account, Townsend's personhood has been erased, her voice literally silenced by cancer. Her flesh is now 'strange', as Ward twice observes; her cancer may be a product of her own physiology, but the growth described is one of an alien substance, which has no concord with the healthy body. The aetiology of Mrs Townsend's cancer was, as in many cases of the disease, troubling and indeterminate. Ward struggled for terms to describe a pathology at once 'cancrous', 'mellicerous' and gristly, which had, for no clear reason, regrown after excision. However unusual it may have been, it is nonetheless clear that this cancer's 'strangeness' was viewed as allied to the strangeness of the female body, and the connection between breast and womb which allowed superfluous and dangerous matter from the latter to accumulate and cause disease in the former. Ward's account does not tell us more specifically about what he, Mrs Townsend or the medical professionals operating on and later dissecting her body believed might have caused her disease. Did Townsend suffer violence, grief or post-natal breast infections, or was her cancer the result of a bad diet and melancholy complexion? Whatever the origin of the disease, it is clear that her symptoms must have been extreme to prompt consent to a mastectomy operation carried out without anaesthetic, in which even the operating surgeons agreed that gangrene and fever were life-threatening possibilities.

This book has analysed medical and non-medical texts in terms of the therapeutic and rhetorical landscape of early modern England, in order to place events like the ones which Ward described into somatic and imaginative context. It is evident that cancer occupied a unique position in the consciousness of not only medical professionals, but lay people and numerous dramatic, persuasive or poetic writers, whether they ever encountered cancerous disease or not. All parties knew cancer as a lethal, cruel and intractable disease. Lay people feared becoming victims of cancer and pitied those whom they saw suffering with the malady. They might have heard of the racking pains inflicted by advanced cancers, or the stinking ulcers which could result from their breaking through the skin. In the face of such gruesome symptoms, it is unsurprising that cancers were widely conceptualised as something apart from and hostile to the body, which ate up one's substance like a ravenous worm or wolf.

Moreover, fear of cancer was not only based upon its morbid physical effects. Early modern bodies were vulnerable to mortal illness and accident in a way that is almost unimaginable to the modern historian, with medicine often largely powerless to stay the spread of infectious disease or assist in a complicated childbirth. Among a wide range of potentially fatal diseases, cancer stood out in part because the malady exceeded the natural body, and was absorbed into the rhetoric of national and institutional sickness. In religious and political polemic, drama, and poetry, the malignancy of cancer came to stand for moral sicknesses concealed beneath an attractive carapace, or for elements or individuals within a group who seemed to belong, but secretly exploited their membership to wreak destruction from the inside. Unsurprisingly, embellishments upon the theme of cancer's evil and cruel 'character' constructed by imaginative writers fed back into the somatic experience of cancerous disease, making cancer a disease of which the medical and literary contexts were inseparable.

Finally, it is worth pointing out, once again, how early modern conceptualisations of cancer may echo into the twenty-first century. The aim of this book has not been to inform modern activist or clinical discourses. Mercifully, much of what is described herein is unrecognizable from modern methods of diagnosis and treatment. Nonetheless, it seems clear that many of the features of today's 'war on cancer' – the adversarial language, the zoomorphic characterisation, the gendering of the disease and its causes – are not, as we may imagine, 'pure' responses to encounters with cancer, but draw on tropes which may be hundreds or even thousands of years old.[10] Twentieth- and twenty-first-century writings about cancer continue to negotiate the same difficult terrain as their sixteenth- and seventeenth-century counterparts. Writing on her own illness and recovery, Hephzibah Roskelly recalls 'bewildered rage at the betrayal by the body', while others identify feelings of de-feminization, or 'occupation' by a foreign entity.[11] While post-Enlightenment discourses may have superficially divided the scientific from the imaginative, cancer still bridges that divide. In both modern and early modern thought, the power of cancer to bring about fear and fascination depends on its status as a powerful traitor: a malady both intimately of the self and, seemingly, ruthlessly hostile toward it.

Notes

Introduction

1. Reverend John Ward, *Diary of the Rev. John Ward, A.M., Extending from 1648 to 1679*, ed. Charles Severn (London: Henry Coldurn, 1839), pp. 245–7.
2. Walter Needham was a Fellow of the Royal College of Physicians and the Royal Society and served as anatomy lecturer to the company of Barber-Surgeons.
3. Ambroise Paré, *The Workes of that Famous Chirurgion Ambrose Parey* (trans. Thomas Johnson, book 29 trans. George Baker) (London: 1634), pp. 280–1.
4. Cancer Research UK, 'Enemy' (advertisement) dir. Siri Bunford, 30 April 2013. This campaign also directly echoes the overdetermination of cancer described by Susan Sontag in the 1970s [*Illness as Metaphor and Aids and Its Metaphors* (London: Penguin, 1991), p. 9].
5. Ellen Leopold. *A Darker Ribbon: Breast Cancer, Women, and Their Doctors in the Twentieth Century* (Boston: Beacon Press, 1999), p. 24.
6. T.D., *The Present State of Chyrurgery, with Some Short Remarks on the Abuses Committed under a Pretence to the Practice. And Reasons Offer'd for Regulating the Same* (London: 1703), pp. 19–20.
7. Doreen Evenden Nagy, *Popular Medicine in Seventeenth-Century England* (Bowling Green, OH: Bowling Green State University Popular Press, 1988); Margaret Pelling (with Frances White), *Medical Conflicts in Early Modern London: Patronage, Physicians, and Irregular Practitioners, 1550–1640* (Oxford: Clarendon Press, 2003); Andrew Wear, *Knowledge and Practice in English Medicine, 1550–1680* (Cambridge: Cambridge University Press, 2000).
8. Wear, *Knowledge and Practice*, p. 23.
9. Ibid.
10. See Seth Stein LeJacq, 'The Bounds of Domestic Healing: Medical Recipes, Storytelling and Surgery in Early Modern England', *Social History of Medicine* 26:3 (2013), 451–68.
11. See Robert Frank, Jr., 'The John Ward Diaries: Mirror of Seventeenth-Century Science and Medicine', *Journal of the History of Medicine* 29 (1974), 147–79.
12. In 1704, the Royal College of Physicians lost their legal monopoly on the practice of physic. The reasons for, and effects of, this loss are discussed at length in Harold Cook, *The Decline of the Old Medical Regime in Stuart London* (New York: Cornell University Press, 1986).
13. Linda Pollock, *With Faith and Physic: The Life of a Tudor Gentlewoman, Lady Grace Mildmay 1552–1620* (London: Collins and Brown Ltd., 1993), p. 97.
14. M.A. Katritzky, *Women, Medicine and Theatre 1500–1750: Literary Mountebacks and Performing Quacks* (Aldershot; Burlington, VT: Ashgate, 2007), p. 125.
15. See Harold Cook, *Trials of an Ordinary Doctor: Joannes Groenevelt in Seventeenth-Century London* (Baltimore: Johns Hopkins University Press, 1994); Philip K. Wilson, *Surgery, Skin and Syphilis: Daniel Turner's London (1667–1741)* (Amsterdam; Atlanta: Rodopi, 1999).

16. See Margaret Pelling, 'Thoroughly Resented? Older Women and the Medical Role in Early Modern London', in Lynette Hunter and Sarah Hutton (eds.), *Women, Science and Medicine 1500–1700: Mothers and Sisters of the Royal Society* (Stroud: Sutton Publishing, 1997), pp. 63–88; Nagy, *Popular Medicine*, especially pp. 55–60.

17. Galen (ed. and trans., with an introduction by Ian Johnston), *Galen: On Diseases and Symptoms* (Cambridge: Cambridge University Press, 2006), especially pp. 10–15.

18. Nancy G. Siraisi, *Medieval and Early Renaissance Medicine: An Introduction to Knowledge and Practice* (Chicago: University of Chicago Press, 1990), p. 193.

19. On William Harvey and new hydraulic models of the body, not discussed here, see Silva De Renzi, 'Old and New Models of the Body', in Peter Elmer (ed.), *The Healing Arts: Health, Disease and Society in Europe, 1500–1800* (Manchester: Manchester University Press, 2004), pp. 166–95; Thomas Wright, *Circulation: William Harvey's Revolutionary Idea* (London: Vintage, 2013).

20. See Roger French and Andrew Wear (eds.), *The Medical Revolution of the Seventeenth Century* (Cambridge: Cambridge University Press, 1989).

21. See Allen G. Debus, *The Chemical Philosophy* (New York: Science History Publications, 1977), p. 107.

22. Lindemann, *Medicine and Society*, p. 102.

23. Jean Baptiste van Helmont, *Van Helmont's Works Containing His Most Excellent Philosophy, Physick, Chirurgery, Anatomy* (trans. 'J.C.') (London: 1664), p. 545.

24. See Charles Webster, 'Paracelsus: Medicine as Popular Protest', in Peter Ole Grell and Andrew Cunningham (eds.), *Medicine and the Reformation* (Abingdon, Oxon: Routledge, 1993), p. 67; Peter Elmer, 'Chemical Medicine and the Challenge to Galenism: The Legacy of Paracelsus, 1560–1700', in Peter Elmer (ed.), *The Healing Arts: Health, Disease and Society in Europe, 1500–1800* (Manchester: Manchester University Press, 2004), p. 128.

25. Mary Lindemann, *Medicine and Society in Early Modern Europe* (2nd edition) (Cambridge; New York: Cambridge University Press, 2010), p. 100.

26. See Silva De Renzi, 'Old and New Models of the Body', in Peter Elmer (ed. and Introduction), *The Healing Arts: Health, Disease and Society in Europe, 1500–1800* (Manchester: Manchester University Press, 2004), p. 181.

27. Lindemann, *Medicine and Society*, p. 87.

28. 'Neo-Galenism' is borrowed from Angus Gowland, *The Worlds of Renaissance Melancholy: Robert Burton in Context* (Cambridge: Cambridge University Press, 2006), p. 40.

29. *Early English Books Online* (Chadwyck), eebo.chadwyck.com; *Eighteenth Century Collections Online* (Gale Group), galegroup.com.lib.exeter.ac.uk/ecco; *Defining Gender, 1450–1910* (Adam Matthew), gender.amdigital.co.uk; Erica Longfellow and Elizabeth Clarke, directors, *Constructing Elizabeth Isham* (University of Warwick), http://www2.warwick.ac.uk/fac/arts/ren/projects/isham.

30. Lynette Hunter, 'Cankers in *Romeo and Juliet*: Sixteenth-Century Medicine at a Figural/Literal Cusp', in Stephanie Moss and Kaara L. Peterson (eds.), *Disease, Diagnosis and Cure on the Early Modern Stage* (Aldershot: Ashgate, 2004), pp. 171–80.

31. Gowland, *The Worlds of Renaissance Melancholy*, p. 49.

32. Marie-Hélène Huet, 'Monstrous Medicine', in Laura Lunger Knoppers and Joan B. Landes (eds.), *Monstrous Bodies/Political Monstrosities in Early Modern Europe* (Ithaca: Cornell University Press, 2004), pp. 127–48.
33. Lesel Dawson, *Lovesickness and Gender in Early Modern English Literature* (Oxford; New York: Oxford University Press, 2008), p. 177.
34. Jan Frans Van Dijkhuizen and Karl A.E. Enenkel (eds.), *The Sense of Suffering: Constructions of Physical Pain in Early Modern Culture* (Leiden; Boston: Brill, 2009), p. 5.
35. See Sarah Covington, *Wounds, Flesh and Metaphor in Seventeenth-Century England* (Basingstoke: Palgrave Macmillan, 2009); William Kerwin, *Beyond the Body: The Boundaries of Medicine and English Renaissance Drama* (Massachusetts: University of Massachusetts Press, 2005); Laura Gowing, *Common Bodies: Women, Touch and Power in Seventeenth-Century England* (New Haven; London: Yale University Press, 2003); Martha Kalnin Diede, *Shakespeare's Knowledgeable Body* (New York: Peter Lang Publishing, Inc., 2008).
36. See Roze Hentschell, 'Luxury and Lechery: Hunting the French Pox in Early Modern England', in Kevin P. Siena (ed.), *Sins of the Flesh: Responding to Sexual Disease in Early Modern Europe* (Toronto: Centre for Reformation and Renaissance Studies, 2005), pp. 133–58; Marie E. McAllister, 'Stories of the Origin of Syphilis in Eighteenth-Century England: Science, Myth, and Prejudice', *Eighteenth-Century Life* 24:1 (2000), 22–44; Rebecca Totaro, *Suffering in Paradise: The Bubonic Plague in Literature from More to Milton* (Pittsburgh, PA: Duquesne University Press, 2005).
37. Barron H. Lerner, *The Breast Cancer Wars: Fear, Hope, and the Pursuit of a Cure in Twentieth-Century America* (New York: Oxford University Press, 2001), especially pp. 269–74; Nadine Ehlers and Shiloh Krupar, 'Introduction: The Body in Breast Cancer', *Social Semiotics* 22:1 (February 2012), 1–11.
38. Ehlers and Krupar, 'Introduction', especially 2.
39. See Leopold, *A Darker Ribbon*, especially pp. 24–6; and James T. Patterson, *The Dread Disease: Cancer and Modern American Culture* (Cambridge, MA: Harvard University Press, 1987), pp. 12–20.
40. Siddhartha Mukherjee, *The Emperor of All Maladies: A Biography of Cancer* (London: Scribner, 2010); James S. Olson, *Bathsheba's Breast: Women, Cancer and History* (Baltimore: Johns Hopkins University Press, 2002); George Johnson, *The Cancer Chronicles: Unlocking Medicine's Deepest Mystery* (London: The Bodley Head, 2013).
41. A. Kaprozilos and N. Pavlidis, 'The Treatment of Cancer in Greek Antiquity', *European Journal of Cancer* 40 (2004), 2033–40.
42. Michael B. Shimkin, *Contrary to Nature: Being an Illustrated Commentary on Some Persons and Events of Historical Importance in the Development of Knowledge Concerning Cancer* (Washington: 1977), pp. 21–2; Carl M. Mansfield, *Early Breast Cancer: Its History and Results of Treatment* (Series: Experimental Biology and Medicine, Vol.5) (Basel; New York: Karger, 1976), p. 2. See also H.S.J. Lee (ed.), *Dates in Oncology* (Carnforth, UK; Pearl River, NY: Parthenon Publishing Group, 2000).
43. See Shimkin, *Contrary to Nature*, p. 42. See also Mukherjee, *The Emperor of All Maladies*, p. 49; Daniel De Moulin, *A Short History of Breast Cancer* (Leiden: Martinus Nijhoff Publishers, 1983), pp. 14–15; Marie-Christine Pouchelle

(trans. Rosemary Morris), *The Body and Surgery in the Middle Ages* (Cambridge: Polity Press, 1990), p. 168.

44. Luke Demaitre, 'Medieval Notions of Cancer: Malignancy and Metaphor', *Bulletin of the History of Medicine* 72:4 (1998), 609–36. See also Luke Demaitre, *Leprosy in Premodern Medicine: A Malady of the Whole Body* (Baltimore: Johns Hopkins University Press, 2007).

45. Demaitre, 'Medieval Notions of Cancer', 620, 622.

46. Wendy D. Churchill, *Female Patients in Early Modern Britain: Gender, Diagnosis, and Treatment* (Farnham: Ashgate, 2012), p. 124; Stolberg, *Experiencing Illness*, pp. 136, 50. See also Mansfield, *Early Breast Cancer*, pp. 4–9.

47. Sujata Iyengar, *Shakespeare's Medical Language: A Dictionary* (London; New York: Continuum, 2011), pp. 51–4.

48. Marjo Kaartinen, *Breast Cancer in the Eighteenth Century* (London; Vermont: Pickering and Chatto, 2013).

49. Sarah Cowper, *Diary* (1700–03) from *Defining Gender* (online resource), http://www.amdigital.co.uk, accessed 13 April 2013. On Cowper's life and diaries, see Faith Lanum, 'Perdita Woman: Sarah Cowper', at *The Perdita Project* (online resource), http://web.warwick.ac.uk/english/perdita.

50. Ibid., p. 71 (19 May 1703), pp. 22–3 (11 November 1700).

51. Judith Butler, *Gender Trouble: Feminism and the Subversion of Identity* (New York: Routledge, 1999).

52. Gowing, *Common Bodies*, p. 4.

53. Robert A. Aronowitz, *Making Sense of Illness: Health, Society, and Disease* (Cambridge: Cambridge University Press, 1998), p. 14; Phil Brown, 'Naming and Framing: The Social Construction of Diagnosis and Illness', *Journal of Health and Social Behavior* 35 (1995), 34–52.

54. Susan Sontag, *Illness as Metaphor and Aids and Its Metaphors* [London: Penguin, 1991 (*Illness as Metaphor* first published 1977, *Aids and Its Metaphors* first published 1988)], p. 3.

55. Ludmilla Jordanova, *Sexual Visions: Images of Gender in Science and Medicine between the Eighteenth and Twentieth Centuries* (Wisconsin: University of Wisconsin Press, 1989), p. 6.

56. Shigehisa Kuriyama, *The Expressiveness of the Body and the Divergence of Greek and Chinese Medicine* (New York; London: Zone Books, 1999), p. 60.

57. See, for example, Mansfield, *Early Breast Cancer*, pp. 1, 2.

58. See Furdell, *Publishing and Medicine*, especially pp. 29–38.

59. The few points of divergence are discussed in De Moulin, *A Short History of Breast Cancer*, pp. 20–30. See also Daniel De Moulin, 'Historical Notes on Breast Cancer, with Emphasis on the Netherlands: I. Pathological and Therapeutic Concepts in the Seventeenth Century', *The Netherlands Journal of Surgery* 32:4 (1980), 129–34; and 'Historical Notes on Breast Cancer, with Emphasis on the Netherlands: II. Pathophysiological Concepts, Diagnosis and Therapy in the 18th Century', *The Netherlands Journal of Surgery* 33:4 (1981), 206–16.

60. See Ole Peter Grell, Andrew Cunningham and Jon Arrizabalaga (eds.), *Centres of Medical Excellence? Medical Travel and Education in Europe, 1500–1789* (Farnham, England; Burlington, Vermont: Ashgate, 2010).

61. Cook, *Trials of an Ordinary Doctor*, pp. 110–11.

62. Carla Mazzio, 'Shakespeare and Science, c. 1600', *South Central Review* 26:1–2 (2009), 1.
63. Furdell, *Publishing and Medicine,* pp. 126–30.
64. See Jennifer Andersen and Elizabeth Sauer (eds.), *Books and Readers in Early Modern England: Material Studies* (Philadelphia: University of Pennsylvania Press, 2002).
65. Ibid., p. 108.
66. Kings-evil was a skin disease, often identical with scrofula, which was supposedly curable by the monarch's touch.
67. Cook, *Trials of an Ordinary Doctor*, p. 113.
68. John Partridge, *The Widowes Treasure* (London: 1588).
69. Doreen Evenden, *The Midwives of Seventeenth-Century London* (Cambridge: Cambridge University Press, 2000), pp. 6–13.
70. These were the English physicians and surgeons, John Pechey, Theodore Mayern, Dr. Chamberlain (probably Thomas Chamberlayne) and Nicholas Culpeper. *The Compleat Midwife's Practice* (London: 1698).
71. Jane Sharp, *The Midwives Book, or, The Whole Art Of Midwifery Discovered* (London: 1671).
72. Alethea Talbot, Countess of Arundel (probable – see Bibliography), *Natura Exenterata* (London: 1655); Hannah Wolley, *The Accomplish'd Ladies Delight in Preserving, Physic, Beautifying, and Cookery* [1686 (1675)].
73. On female-authored almanacs, see A.S. Weber, 'Women's Early Modern Medical Almanacs in Historical Context', *English Literary Renaissance* 33:3 (November 2003), 358–402.
74. Elaine Leong and Sara Pennell, 'Recipe Collections and the Currency of Medical Knowledge in the Early Modern "Medical Marketplace"', in Mark S.R Jenner and Patrick Wallis (eds.), *Medicine and the Market in England and its Colonies, c.1450–c.1850* (Basingstoke, NY: Palgrave Macmillan, 2007), pp. 133–53.
75. All the casebooks used in this work are male-authored, although some female medical practitioners, especially midwives, kept similar records. See Evenden, *The Midwives of Seventeenth-Century London*, p. 128.
76. Richard Wiseman, *Several Chirurgical Treatises* (second edition) (London: 1686); John Hall, *Select Observations on English Bodies* (1657).

1 What Was Cancer? Definition, Diagnosis and Cause

1. Nathan Bailey, *An Universal Etymological English Dictionary* (London: 1721), sig. R3v.
2. See Daniel De Moulin, *A Short History of Breast Cancer* (Leiden: Martinus Nijhoff Publishers, 1983), p. 20; George H. Sakorafas and Michael Safioleas, 'Breast Cancer Surgery: An Historical Narrative. Part I. From Prehistoric Times to Renaissance', *European Journal of Cancer Care* 18:6 (November 2009), 540.
3. Marjo Kaartinen, *Breast Cancer in the Eighteenth Century* (London; Vermont: Pickering and Chatto, 2013), pp. 2–7.
4. Jonathan Gil Harris, '"The Canker of England's Commonwealth": Gerard Malynes and the Origins of Economic Pathology', *Textual Practice* 13:2 (1999), 311–28.

5. Lynette Hunter, 'Cankers in *Romeo and Juliet*: Sixteenth-Century Medicine at a Figural/Literal Cusp', in Stephanie Moss and Kaara L. Peterson (eds), *Disease, Diagnosis and Cure on the Early Modern Stage* (Aldershot: Ashgate, 2004), p. 171.

6. Ibid.

7. Sujata Iyengar, *Shakespeare's Medical Language: A Dictionary* (London; New York: Continuum, 2011), p. 52.

8. Pauline Thompson, 'The Disease That We Call Cancer', in S. Campbell, B. Hall and D. Klausner (eds), *Health, Disease and Healing in Medieval Culture* (Basingstoke: Palgrave Macmillan, 1992), p. 2.

9. Luke Demaitre, 'Medieval Notions of Cancer: Malignancy and Metaphor', *Bulletin of the History of Medicine* 72:4 (1998), 623.

10. Luke Demaitre, *Leprosy in Premodern Medicine: A Malady of the Whole Body* (Baltimore: Johns Hopkins University Press, 2007), p. 91.

11. Théophile Bonet, *A Guide to the Practical Physician* (London: 1684), p. 62.

12. R.W. McConchie, *Lexicography and Physicke: the Record of Sixteenth-Century English Medical Terminology* (Oxford; New York: Oxford University Press, 1997), p. 204 (author's emphases).

13. Pierre Dionis, *A Course of Chirurgical Operations, Demonstrated in the Royal Garden at Paris* [London: 1710 (French edition 1707)], pp. 247–8. See also John Browne, *The Surgeons Assistant* (London: 1703), p. 84; Giovannida Vigo, *The Most Excellent Workes of Chirurgerie* [London: 1571 (1543)], p. xliv; John Smith, *A Compleat Practice of Physic* (London: 1656), p. 52.

14. See Demaitre, who finds noli-me-tangere to have been identified as a 'subspecies' of cancer by medieval medical practitioners (Demaitre, 'Medieval Notions of Cancer', 616).

15. Matthias Gottfried Purmann (with appended text by Conrade Joachim Sprengell), *Chirurgia Curiosa .. To Which is Added Natura Morborum Medicatrix: Nature Cures Diseases* (London: 1706), p. 34.

16. George Wither, 'Opobalsamum Anglicanum', in *Miscellaneous Works* [1872–1877 (c.1645)], p. 149. From *English Poetry Database* (online resource), www.0-collections.chadwyck.co.uk, 19 February 2011.

17. Philip Barrough, *The Method of Physick Conteyning the Causes, Signes, and Cures of Inward Diseases in Mans Body From the Head to the Foote* (London: 1583), p. 274. See also: James Handley, *Colloquia Chirurgica: Or, the Whole Art of Surgery Epitomiz'd and Made Easie* (London: 1705), p. 66; Alexander Read, *The Workes of That Famous Physician Dr. Alexander Read* (second edition) (London: 1650), p. 171.

18. Richard Wiseman, *Several Chirurgical Treatises* [London: 1686 (first edition 1676)], p. 99.

19. Ibid., p. 101.

20. John Pechey and Theodore Mayern (Sir Théodore Turquet de Mayerne), 'Dr. Chamberlain', Nicholas Culpeper, *The Compleat Midwife's Practice* (London: 1698), pp. 183–4. See the Bibliography for more on the provenance of this text. See also: Ambroise Paré, *The Workes* (trans. Thomas Johnson, book 29 trans. George Baker) [London: 1634 (collated from 16th-century texts)], p. 279; Dionis, *A Course of Chirurgical Operations*, p. 248; Wiseman *Several Chirurgical Treatises*, p. 98; Read, *The Chirurgicall Lectures*, pp. 211, 213–14; Browne, *The Surgeons Assistant*, p. 80; Peter Lowe, *The Whole Course of*

Chirurgerie ... Whereunto is Annexed the Presages of Divine Hippocrates (London: 1597), sig. L3r; John Pechey, *The Store-House of Physical Practice* (London: 1695), p. 61; *An Account of the Causes of some Particular Rebellious Distempers viz. the Scurvey, Cancers in Women's Breasts, &c. Vapours, and Melancholy, &c. Weaknesses in Women, &c. Gout, Fistula in Ano, Dropsy, Agues, &c.* (London: 1670), p. 23; Vigo, *The Most Excellent Workes of Chirurgerie*, p. xliii; Gendron, *Enquiries*, p. 54.

21. Paré, *The Workes*, p. 148; See also Robert Bayfield, *Tractatus de Tumoribus Praeter Naturam, or, A Treatise of Preternatural Tumors* (London: 1662), p. 180; Dionis, *A Course of Chirugical Operations*, p. 248; Wiseman, *Several Chirurgical Treatises*, p. 98; Read, *The Chirurgicall Lectures*, p. 211; Barrough, *The Method of Physick*, p. 273; Paul Dubé, *The Poor Man's Physician and Surgeon* (eighth edition) (London: 1704), p. 362; John Tanner, *The Hidden Treasures of the Art of Physick* (1659), p. 441.

22. Jacques Guillemeau, 'A.H.', and W. Bailey, *A Worthy Treatise of the Eyes ... Togeather With a Profitable Treatise of the Scorbie; & Another of the Cancer by A. H.* (London: 1587), p. 40. See also for examples: Bayfield, *Tractatus de Tumoribus*, p. 180; Nicholas Culpeper, *A Directory for Midwives, or, A Guide for Women, in their Conception, Bearing, and Suckling their Children* (London: 1651), p. 324; Wiseman, *Several Chirurgical Treatises*, p. 98; Maynwaringe, *The Frequent, but Unsuspected Progress of Pains*, pp. 194–5; John Moyle, *The Experienced Chirurgion* (London: 1703), p. 48; Dubé, *The Poor Man's Physician* p. 362; William Salmon, *Paraieremata, or Select Physical and Chirurgical Observations* (London: 1687), pp. 377–8; Paul Barbette with Raymundus Minderius (Raymond Minderer), *Thesaurus Chirurgiae: The Chirurgical and Anatomical Works of Paul Barbette* (London: 1687 [1676]), pp. 122–3.

23. Pechey et al., *The Compleat Midwife's Practice*, p. 183.

24. Demaitre, 'Medieval Notions of Cancer', 612.

25. Barrough, *The Method of Physick*, p. 144. See also Bayfield, *Tractatus de Tumoribus*, p. 180; Wiseman, *Several Chirurgical Treatises*, p. 102; Giovanni Lanfranco (Lanfranco of Milan), *A Most Excellent and Learned Woorke of Chirurgerie, called Chirurgia parua Lanfranci* (trans. John Halle) (London: 1565), p. 20; John Browne, *Adenochoiradelogia, or, An Anatomick-Chirurgical Treatise of Glandules & Strumaes or, Kings-Evil-Swellings* (London: 1684), pp. 31–2.

26. One notable exception to this rule was Culpeper, who argued that cancers were painless until they grew large or ulcerated (Culpeper, *A Directory for Midwives*, p. 165).

27. Christof Wirsung (trans. Jacob Mosan), *Praxis Medicinae Universalis* (London: 1598), p. 572.

28. Wiseman, *Several Chirurgical Treatises*, p. 98; Browne, *Adenochoiradelogia*, pp. 31–2. See also Barrough, *The Method of Physick*, p. 274; Henri-François Le Dran, *Observations in Surgery* (trans. John Sparrow) (London: 1739), p. 156.

29. Barrough, *The Method of Physick*, p. 274. See also Dubé, *The Poor Man's Physician*, p. 362; Wirsung, *Praxis Medicinae Universalis*, p. 498.

30. Paré, *The Workes*, p. 279.

31. Lowe, *The whole course of chirurgerie*, sig.L 3r.

32. Barrough, *The Method of Physick*, p. 273; Read, *The Chirurgicall Lectures*, pp. 211–12.

33. Dionis, *A Course of Chirurgical Operations*, p. 248; Pare, *The Workes*, p. 279.

34. Browne, *The Surgeons Assistant*, p. 81.
35. Maynwaringe, *The Frequent, but Unsuspected Progress of Pains*, p. 183.
36. Ibid., p. 188.
37. See Bridget Gellert Lyons, *Voices of Melancholy: Studies in Literary Treatments of Melancholy in Renaissance England* (London: Routledge & Kegan Paul, 1971), p. 2; Robert Burton, *The Anatomy of Melancholy* (Oxford: 1621), p. 21.
38. Burton, *The Anatomy of Melancholy*, p.78.
39. Ibid.
40. Bayfield, *Tractatus de Tumoribus*, pp. 92–3.
41. Barrough, *The Method of Physick*, p. 273, see also p. 276; Browne, *The Surgeons Assistant*, p. 81. On the related belief that strong wine heated the body and exacerbated cancers, see Chapter 5.
42. Paré, *The Workes*, pp. 279–80. See also Browne, *The Surgeons Assistant*, p. 82; Pechey et al., *The Compleat Midwife's Practice*, p. 184; Handley, *Colloquia Chirurgica*, p. 68; Read, *The Workes*, p. 170.
43. Read, *Chirurgicall Lectures*, p. 212.
44. Demaitre, 'Medieval Notions of Cancer', 619.
45. Angus Gowland, 'The Problem of Early Modern Melancholy', *Past and Present* 191 (May 2006), 86–7; see also Demaitre, 'Medieval Notions of Cancer', 618.
46. Gowland, 'The Problem of Early Modern Melancholy', 82.
47. Charles Taylor, *Sources of the Self: The Making of the Modern Identity* (Cambridge, MA: Cambridge University Press, 1989), pp. 188–9.
48. Burton, *Anatomy of Melancholy*, p. 45.
49. Gowland, 'The Problem of Early Modern Melancholy', 92. See also Gowland, *The Worlds of Renaissance Melancholy*, p. 86. Erica Fudge notes that melancholy madness was thought, from the seventeenth century, to account for stories of lycanthropia (werewolves) [Erica Fudge, *Perceiving Animals: Humans and Beasts in Early Modern English Culture* (Basingstoke: Macmillan Press Ltd., 2000), p. 54].
50. Clark, *Vanities of the Eye*, p. 52.
51. Lyons, *Voices of Melancholy*, pp. 4–5.
52. Burton, *The Anatomy of Melancholy*, p. 21.
53. Gail Kern Paster, 'Melancholy Cats, Lugged Bears, and Early Modern Cosmology: Reading Shakespeare's Psychological Materialism Across the Species Barrier', in Gail Kern Paster, Katherine Rowe and Mary Floyd-Wilson (eds), *Reading the Early Modern Passions: Essays in the Cultural History of Emotion* (Philadelphia: University of Pennsylvania Press, 2004), p. 118.
54. Browne, *The Surgeons Assistant*, p. 81; Read, *Chirurgicall Lectures*, pp. 214–15.
55. Gendron, *Enquiries*, p. 5; Wiseman, *Several Chirurgical Treatises*, pp. 98–9.
56. Read, *Chirurgicall Lectures*, p. 212.
57. Lyons, *Voices of Melancholy*, p. 2. See also Gowland, *The Worlds of Renaissance Melancholy*, p. 63.
58. Significantly, many dietary prescriptions were designed to reduce choler – see Chapter 5.
59. Read, *Chirurgicall Lectures*, p. 212; Vigo, *The Most Excellent Workes*, p. xliii. See also Browne, *The Surgeons Assistant*, p. 76; Barrough, *The Method of Physick*, p. 202.
60. Browne, *Adenochoiradelogia*, p. 31–2.

61. Katharine A. Craik, *Reading Sensations in Early Modern England* (Basingstoke: Palgrave Macmillan, 2007), p. 57; Burton, *The Anatomy of Melancholy*, p. 51.
62. Jennifer Radden, *The Nature of Melancholy: From Aristotle to Kristeva* (Oxford: Oxford University Press, 2002), p. 63.
63. See *An Account of the Causes of Some Particular Rebellious Distempers*, p. 21–3; Moyle, *The Experienced Chirurgion*, p. 48.
64. See, for example, Andrew Wear, *Knowledge and Practice in English Medicine, 1550–1680* (Cambridge: Cambridge University Press, 2000), especially Chapter 8, 'Conflict and Revolution in Medicine – the Helmontians', pp. 353–98; Peter Elmer, 'Chemical Medicine and the Challenge to Galenism: The Legacy of Paracelsus, 1560–1700', in Peter Elmer (ed.), *The Healing Arts: Health, Disease and Society in Europe, 1500–1800* (Manchester: Manchester University Press, 2004), pp. 108–35.
65. Wiseman, *Several Chirurgical Treatises*, p. 98.
66. Ibid., pp. 98–9; Wiseman, *Several Chirurgical Treatises*, p. 99.
67. Gendron, *Enquiries*, p. 6.
68. Ibid., p. 8. On Gendron's theory of cancerous growth, see also Chapter 4.
69. Ibid., pp. 25–6.
70. Ibid., pp. 66, 67.
71. Wiseman, *Several Chirurgical Treatises*, pp. 101–2.
72. Paster, 'Melancholy Cats, Lugged Bears and Early Modern Cosmology', p. 118.
73. Samuel Butler, 'Religion', from *Satires and Miscellaneous Poetry and Prose* (1928: the date of writing is unknown, although Butler was most active from 1650 to 1680), l.51–54. From *English Poetry Database* (online resource), www.0-collections.chadwyck.co.uk, 7 February 2011.
74. Paul Dubé, *The Poor Man's Physician and Surgeon* (London: 1704), p. 333.
75. Browne, *The Surgeons Assistant*, p. 81.
76. Bonet, *A Guide to the Practical Physician*, p. 62.

2 Cancer and the Gendered Body

1. Sarah Cowper, *Diary* (1700–03). Accessed via *Defining Gender* (online resource), http://www.amdigital.co.uk/m-collections/collection/defining-gender-1450–1910/, 5 June 2009.
2. Thomas Laqueur, *Making Sex: The Body and Gender from the Greeks to Freud* (Cambridge, MA; London: Harvard University Press, 1990); Londa Schiebinger, 'Skeletons in the Closet: The First Illustrations of the Female Skeleton in Eighteenth-Century Anatomy', *Representations* 14, 'The Making of the Modern Body: Sexuality and Society in the Nineteenth Century', (1986), 42–82.
3. Michael Stolberg, 'A Woman Down to Her Bones: The Anatomy of Sexual Difference in the Sixteenth and Early Seventeenth Centuries', *Isis* 94:2 (2003), 274–99.
4. Ibid., 276.
5. Ibid., 299.
6. Gail Kern Paster, 'The Unbearable Coldness of Female Being: Women's Imperfection and the Humoral Economy', *English Literary Renaissance* 28 (1998), 416–40.

7. Daniel De Moulin, 'Historical Notes on Breast Cancer, with Emphasis on the Netherlands: I. Pathological and Therapeutic Concepts in the Seventeenth Century', *The Netherlands Journal of Surgery* 32:4 (1980), 129.
8. James S. Olson, *Bathsheba's Breast: Women, Cancer, and History* (Baltimore, MA: Johns Hopkins University Press, 2002), p. 295.
9. See, for example, Alexander Read, *Most Excellent and Approved Medicine* (London: 1651), p. 100; Owen Wood, *An Alphabeticall Book of Physical Secrets* (London: 1639), p. 32.
10. John Marten, *Gonosologium Novum: Or, a New System of all the Secret Infirm and Diseases, Natural, Accidental, and Venereal in Men and Women* (London: 1709), p. 31.
11. Ibid., pp. 31–2.
12. On the 'one-sex' model, see Laqueur, *Making Sex*, especially Chapter 3, 'New Science, One Flesh', pp. 59–109.
13. John Browne, *The Surgeons Assistant.. Also a Compleat Treatise of Cancers and Gangreens. With an Enquiry Whether they have any Alliance with Contagious Diseases* (London: 1703), pp. 109–10.
14. Robert Bayfield, *Enchiridion Medicum* (London: 1655), pp. 293–4.
15. *An Account of the Causes of Some Particular Rebellious Distempers viz. the Scurvey, Cancers in Women's Breasts, &c. Vapours, and Melancholy, &c. Weaknesses in Women, &c. Gout, Fistula in Ano, Dropsy, Agues, &c.* (London: 1670), p. 24. Second italics my own.
16. Marjo Kaartinen, *Breast Cancer in the Eighteenth Century* (London; Vermont: Pickering and Chatto, 2013), p. 8; Luke Demaitre, 'Medieval Notions of Cancer: Malignancy and Metaphor', *Bulletin of the History of Medicine* 72:4 (1998), 610; James Handley, *Colloquia Chirurgica: Or, the Whole Art of Surgery Epitomiz'd and Made Easie* (London: 1705), p. 66.
17. Nathan Bailey, *An Universal Etymological English Dictionary* (London: 1721), sig. R3v.
18. Edward Shorter, *A History of Women's Bodies* (Harmondsworth: Pelican, 1984), p. 242.
19. Thomas Adams, *The Blacke Devil or the Apostate. Together with the Wolfe Worrying the Lambes and The Spirituall Navigator, Bound for the Holy Land* (London: 1615), pp. 31–2.
20. John Webster, *The White Devil* (1612), in René Weis (ed.), *The Duchess of Malfi and Other Plays* (Oxford: Oxford University Press, 1996), pp. 1–103. On cankers in Shakespeare's Sonnets, see my chapter 'The Worm and the Flesh: Cankered Bodies in Shakespeare's Sonnets', in Sujata Iyengar (ed.), *Disability, Health and Happiness in the Shakespearean Body* (New York; Abingdon, UK: Routledge, 2015), pp. 240–60.
21. Katherine Park, *Secrets of Women: Gender, Generation and the Origins of Human Dissection* (New York: Zone Books, 2010), p. 26. See also Robert Martensen, 'The Transformation of Eve: Women's Bodies, Medicine and Culture in Early Modern England', in Roy Porter and Mikulas Teich (eds), *Sexual Knowledge, Sexual Science: The History of Attitudes to Sexuality* (Cambridge: Cambridge University Press, 1994), pp. 107–33.
22. Matthew Cobb, *The Egg and Sperm Race: The Seventeenth-Century Scientists Who Unravelled the Secrets of Sex, Life and Growth* (London: The Free Press, 2006);

Monica Green, 'From Diseases of Women to Secrets of Women', *Journal of Medieval and Early Modern Studies* 30 (2000), 5–39.

23. Michael Stolberg, 'Menstruation and Sexual Difference in Early Modern Medicine', in Andrew Shail and Gillian Howie (eds), *Menstruation: A Cultural History* (London: Palgrave Macmillan, 2005), pp. 90–101.

24. Ibid. See also Gianna Pomata, 'Menstruating Men: Similarity and Difference of the Sexes in Early Modern Medicine', in Valeria Finucci and Kevin Brownlee (eds), *Generation and Degeneration: Tropes of Reproduction in Literature and History from Antiquity to Early Modern Europe* (Durham, NC; London: Duke University Press, 2001), pp. 108–52.

25. Gail Kern Paster, *The Body Embarrassed: Drama and the Disciplines of Shame in Early Modern England* (New York: Cornell University Press, 1993), especially Chapter 2, 'Laudable Blood: Bleeding, Difference, and Humoral Embarrassment', pp. 64–112.

26. Stolberg, 'Menstruation and Sexual Difference', pp. 91–2.

27. See Monica H. Green, 'Flowers, Poisons and Men: Menstruation in Medieval Western Europe', in Andrew Shail and Gillian Howie (eds), *Menstruation: A Cultural History* (London: Palgrave Macmillan, 2005), pp. 51–64.

28. John Sadler, *The Sicke Woman's Private Looking-Glasse. Wherein Methodically are Handled all Uterine Affects or Diseases Arising from the Womb. Enabling Women to Informe the Physitian About the Cause of their Griefe* (London: 1636), sig. A5r.

29. Jean Riolan, *A Sure Guide, or, The Best and Nearest Way to Physick and Chyrurgery* (trans. Nicholas Culpeper and W.R.) (London: 1657), p. 85.

30. Thomas Bartholin (published, with possible additions, by Nicholas Culpeper and Abdiah Cole), *Bartholinus Anatomy* (London: 1668), p. 70.

31. Ibid.

32. Nicholas Culpeper, *A Directory for Midwives, or, A Guide for Women, in their Conception, Bearing, and Suckling their Children* (London: 1651), pp. 165–6.

33. Lazarius Riverius, *The Practice of Physick* (trans. with additions by Nicholas Culpeper, Abdiah Cole and William Rowland) (London: 1655), p. 492.

34. Sir Edmund King, *Sir Edmund King's Casebook, 1676–96*, British Library, Sloane MS.1589, p. 297v (pagination is irregular).

35. Ibid.

36. Robert Bayfield, *Tractatus de Tumoribus Praeter Naturam, or, A Treatise of Preternatural Tumors* (London: 1662), p. 190; Paré, *The Workes*, p. 282.

37. Christof Wirsung, *Praxis Medicinae Universalis* (trans. Jacob Mosan) (London: 1598), p. 498; Alexander Read, *The Chirurgicall Lectures of Tumors and Ulcers* (London: 1635), p. 215.

38. See, for example, Shakespeare's 'Sonnet 35', 'Sonnet 70', 'Sonnet 95', all of which describe an entity of 'flower' which appears lovely but is inwardly consumed by moral or social 'cankers'.

39. Riverius, *The Practice of Physick*, p. 492. On debates over the nature of menstrual blood, see Paster, *The Body Embarrassed*, pp. 79–81.

40. See, for example, Paré, *The Works*, p. 280; *An Account of the Causes of Some Particular Rebellious Distempers*, p. 22; Jacques Guillemeau, 'A.H.' and W. Bailey, *A Worthy Treatise of the Eyes .. Togeather With a Profitable Treatise of the Scorbie; & Another of the Cancer by A. H.* (London: 1587), p. 46.

41. Pierre Dionis, *A Course of Chirurgical Operations, Demonstrated in the Royal Garden at Paris* (London: 1710 [French edition 1707]) p. 249. See also Claude Deshaies Gendron, *Enquiries into the Nature, Knowledge, and Cure of Cancers* (London: 1701), p. 33.
42. Michael Stolberg, 'A Woman's Hell? Medical Perceptions of Menopause in Pre-Industrial Europe', *Bulletin of the History of Medicine* 73 (1999), 408–28. Churchill also points out that while menopause was not generally deemed pathological, physicians often treated irregular menstruation in women who were entering the menopause in the same manner as amenorrhea in younger women (Churchill, *Female Patients*, p. 114).
43. Elizabeth Lane Furdell, *The Royal Doctors, 1485–1714: Medical Personnel at the Tudor and Stuart Courts* (New York: University of Rochester Press, 2001), p. 56. See also Cathy McClive, 'The Hidden Truths of the Belly: The Uncertainties of Pregnancy in Early Modern Europe', *Social History of Medicine* 15:2 (2002), 221.
44. Galen of Pergamon, *Certaine Workes of Galens, Called Methodus Medendi*, trans. Thomas Gale (London: 1566), p. 365.
45. *An Account of the Causes of Some Particular Rebellious Distempers*, p. 24.
46. Philip Barrough, *The Method of Physick* (London: 1583), p. 145.
47. See Churchill, *Female Patients*, pp. 125–6.
48. See Laqueur, *Making Sex*, pp. 104–5.
49. Riolan, *A Sure Guide*, p. 98
50. Browne, *The Surgeons Assistant*, p. 77.
51. *An Account of the Causes of Some Particular Rebellious Distempers*, p. 22.
52. See Patricia Crawford, 'Attitudes to Menstruation in Seventeenth-Century England', *Past and Present* 91 (1981), 50–52; Green, 'Flowers, Poisons and Men', p. 54; Barbara Orland, 'White Blood and Red Milk: Analogical Reading in Medical Practice and Experimental Physiology (1560–1730)', in Manfred Horstmanshoff, Helen King and Claus Zittel (eds), *Blood, Sweat and Tears: The Changing Concepts of Physiology from Antiquity into Early Modern Europe* (Leiden: Brill, 2012), pp. 443–78.
53. Sadler, *The Sicke Woman's Private Looking-Glasse*, pp. 10–11.
54. *An Account of the Causes of Some Particular Rebellious Distempers*, pp. 22–3.
55. Browne, *The Surgeons Assistant*, p. 77.
56. Reverend John Ward, *Diary of the Rev. John Ward, A.M., Extending from 1648 to 1679*, ed. Charles Severn (London: Henry Coldurn, 1839), p. 247. From *Internet Archive* (online resource), http://www.archive.org, 2 March 2012.
57. *The Compleat Doctoress: Or, a Choice Treatise of all Diseases Insident to Women* (London: 1656), p. 45.
58. Riolan, *A Sure Guide*, p. 97.
59. John Pechey, Theodore Mayern (Sir Théodore Turquet de Mayerne), Dr. Chamberlain (probably Thomas Chamberlayne) and Nicholas Culpeper, *The Compleat Midwife's Practice* (London: 1698), p. 186; Paster, *The Body Embarrassed*, p. 205.
60. Johannes Jonstonus, *The Idea of Practicall Physick* (trans. Nicholas Culpeper) (London: 1657), p. 25.
61. Pechey et al., *The Compleat Midwife's Practice*, p. 186.

62. On the character of the 'ideal' man, see Alexandra Shepard, *Meanings of Manhood in Early Modern England* (Oxford: Oxford University Press, Inc., 2003), especially Chapter 2, 'The Imagined Body of "Man's Estate"', pp. 47–69.
63. See Sarah Toulalan, '"To[o] Much Eating Stifles the Child": Fat Bodies and Reproduction in Early Modern England', *Historical Research* 87:235 (2014), especially 77–9.
64. Riolan, *A Sure Guide*, p. 97. See also Jane Sharp, *The Midwives Book* (London: 1671), p. 360.
65. Angela McShane Jones, 'Revealing Mary', *History Today* (March, 2004), 40–6; Paster, *The Body Embarrassed*, pp. 204–5.
66. Marilyn Yalom, *A History of the Breast* (London: Pandora, 1998), p. 74.
67. Thomas Gibson, *The Anatomy of Humane Bodies Epitomized* (London: 1682), p. 211.
68. Felicity Nussbaum, *Torrid Zones: Maternity, Sexuality, and Empire in Eighteenth-Century English Narratives* (Baltimore; London: Johns Hopkins University Press, 1995), p. 111; Sarah Toulalan, *Imagining Sex: Pornography and Bodies in Seventeenth-Century England* (Oxford: Oxford University Press, 2007), p. 79.
69. McShane Jones, 'Revealing Mary', 44.
70. Margaret R. Miles, *A Complex Delight: The Secularization of the Breast 1350–1750* (Berkeley: University of California Press, 2008), p. 2.
71. Valerie Fildes, *Breasts, Bottles and Babies: A History of Infant Feeding* (Edinburgh: Edinburgh University Press, 1986), p. 98; David Harley, 'From Providence to Nature: The Moral Theology and Godly Practice of Maternal Breast-feeding in Stuart England', *Bulletin of the History of Medicine* 69:2 (1995), 198–223.
72. Jacques Guillemeau, *Childbirth, or, the Happy Delivery of Women* (London: 1612), sig. Ii2v.
73. *An Account of the Causes of Some Particular Rebellious Distempers*, p. 20; Riolan, *A Sure Guide*, p. 98; Pechey, *The Compleat Midwife's Practice*, p. 164.
74. Sharp, *The Midwives Book*, p. 338.
75. This was conspicuously the case in some nineteenth-century texts; most notably, Maria Edgeworth's novel *Belinda*, in which Lady Delacour believes (incorrectly) that she has breast cancer and ascribes this to her failure to breastfeed her children [Maria Edgeworth, *Belinda* (1801) (New York: Oxford University Press, 1996)].
76. Pierre Dionis, *A General Treatise of Midwifery* (London: 1719), p. 292.
77. Shorter, *A History of Women's Bodies*, p. 244.
78. *Collection of Medical and Cookery Receipts* (early seventeenth century), Wellcome Library, MS.635, p. 24.
79. Sharp, *The Midwives Book*, pp. 338–47; Culpeper, *A Directory for Midwives*, pp. 324–6; Riolan, *A Sure Guide*, p. 98.
80. Lawrence Stone, *The Family, Sex and Marriage in England, 1500–1800* (New York: Harper & Row, 1977); Susan Amussen, *An Ordered Society: Gender and Class in Early Modern England* (New York: Columbia University Press, 1988); Su Fang Ng, *Literature and the Politics of Family in Seventeenth-Century England* (Cambridge: Cambridge University Press, 2007).
81. Dionis, *A Course of Chirurgicall Operations*, p. 249.
82. Madame de Motteville, *Memoirs of Madame de Motteville on Anne of Austria and Her Court* (trans. Katherine Prescott Wormeley, with an introduction by C.A. Sainte-Beuve) (Boston: Hardy, Pratt & Company, 1902), pp. 310, 186.

83. See Miles, *A Complex Delight*, p. 110.
84. Sarah E. Owens, 'The Cloister as Therapeutic Space: Breast Cancer Narratives in the Early Modern World', *Literature and Medicine* 30:2 (2012), 322. Original quotation from Wilmer Cave Wright (ed.), *De Morbis Artificum by Barnardino Ramazzini: The Latin Text of 1713* (London: 1940), p. 191.
85. On menstrual disorders and nuns, see Green, 'Flowers, Poisons and Men', p. 56; Cathy McClive, 'Menstrual Knowledge and Medical Practice in Early Modern France, c.1555–1761', in Andrew Shail and Gillian Howie (eds), *Menstruation: A Cultural History* (London: Palgrave Macmillan, 2005), pp. 76–89. Carol Thomas Neely notes that hysteria, a disease with a number of pathological similarities to cancer, was thought mostly to affect virgins, nuns and widows; see '"Documents in Madness": Reading Madness and Gender in Shakespeare's Tragedies and Early Modern Culture', *Shakespeare Quarterly* 42:3 (Autumn 1991), 320. Lovesickness and greensickness, associated in part with humoral flux, were also most likely to affect the unmarried woman and could often be 'cured' with sexual intercourse [Lesel Dawson, *Lovesickness and Gender in Early Modern English Literature* (Oxford; New York: Oxford University Press, 2008), especially pp. 47–9].
86. According to Kaartinen, 'mechanical causes' remained prominent into the nineteenth century, though she does not identify them as referring to violence (*Breast Cancer in the Eighteenth Century*, pp. 17–18).
87. Sharp, *The Midwives Book*, p. 339; Gendron, *Enquiries*, p. 38; Barrough, *The Method of Physick*, p. 207; *An Account of the Causes of Some Particular Rebellious Distempers*, p. 21.
88. *Daily Journal* Issue 2652 (London: 8 July 1729). From *Burney Newspaper Collections* (online resource), http://0-find.galegroup.com, 4 March 2013.
89. *An Account of the Causes of Some Particular Rebellious Distempers*, p. 23.
90. Ibid., p. 29.
91. Ibid., p. 30.
92. Roy Porter, *Bodies Politic: Disease, Death and Doctors in Britain, 1650–1900* (London: Reaktion Books, 2001), p. 36.
93. Barrough, *The Method of Physick*, p. 207.
94. Garthine Walker, *Crime, Gender, and Social Order in Early Modern England* (Cambridge; New York: Cambridge University Press, 2003), pp. 63–70; Elizabeth Foyster, 'A Laughing Matter? Marital Discord and Gender Control in Early Modern England', *Rural History* 4:1 (1993), 5–21; Laura Gowing, *Domestic Dangers: Women, Words and Sex in Early Modern London* (Oxford: Clarendon Press, 1996), especially Chapter 6, 'Domestic Disorders: Adultery and Violence', pp. 180–231.
95. Gowing, *Domestic Dangers*, p. 206.
96. Churchill draws the same conclusion in relation to medical texts' silence on traumatic injuries to women more generally (*Female Patients*, p. 50).
97. Cowper, *Diaries*, p. 64.
98. *An Account of the Causes of Some Particular Rebellious Distempers*, p. 22; Read, *The Works*, pp. 172–3; Paré, *The Workes*, p. 280.
99. Evelyne Berriot-Salvadore, 'The Discourse of Medicine and Science' (trans. Arthur Goldhammer), in Natalie Zemon Davis and Arlette Farge (eds), *A History of Women: Renaissance and Enlightenment Paradoxes* (Cambridge, MA: Belknap Press of Harvard University Press, 1993), p. 354.

100. Gowing, *Domestic Dangers*, pp. 208–14.
101. Cowper, *Diaries*, pp. 22–3.
102. Gowing, *Domestic Dangers*, p. 211.
103. Hephzibah Roskelly, 'I meditate on Descartes', *Social Semiotics* 22:1 (2012), 35.

3 'It Is, Say Some, of a Ravenous Nature': Zoomorphic Images of Cancer

Daniel Turner, *De Morbis Cutaneis: Diseases Incident to the Skin* (London: 1714), p. 75.

1. Keith Thomas, *Man and the Natural World: Changing Attitudes in England, 1500–1800* (London: Penguin, 1983); Margaret Healy, 'Bodily Regimen and Fear of the Beast: "Plausibility" in Renaissance Domestic Tragedy', in Erica Fudge, Ruth Gilbert and Susan Wiseman (eds), *At the Borders of the Human: Beasts, Bodies and Natural Philosophy in the Early Modern Period* (Basingstoke: Macmillan, 1999), pp. 51–73. See also Erica Fudge, *Perceiving Animals: Humans and Beasts in Early Modern English Culture* (Basingstoke: Macmillan Press Ltd, 2000).
2. Jean E. Feerick and Vin Nardizzi (eds), *The Indistinct Human in Renaissance Literature* (Basingstoke: Palgrave Macmillan, 2012).
3. Ian MacInnes, 'The Politic Worm: Invertebrate Life in the Early Modern English Body', in Jean E. Feerick and Vin Nardizzi (eds), *The Indistinct Human in Renaissance Literature* (Basingstoke: Palgrave Macmillan, 2012), pp. 253–74.
4. Karen Edwards, 'Milton's Reformed Animals: An Early Modern Bestiary' series, published in instalments in *Milton Quarterly* 39:3 to 43:4 (2005–2009). See especially 'Milton's Reformed Animals: An Early Modern Bestiary A–C', *Milton Quarterly* 39:4 (2005), 183–292; and 'Milton's Reformed Animals: An Early Modern Bestiary T–Z', *Milton Quarterly* 43:4 (2009), 241–303.
5. Marta Powell Harley, 'Last Things First in Chaucer's Physician's Tale: Final Judgment and the Worm of Conscience', *The Journal of English and Germanic Philology* 91:1 (1992), 1–16; Jonathan Wright, 'The World's Worst Worm: Conscience and Conformity During the English Reformation', *The Sixteenth Century Journal* 30:1 (1999), 113–33.
6. Jonathan Gil Harris, '"The Canker of England's Commonwealth": Gerard Malynes and the Origins of Economic Pathology', *Textual Practice* 13:2 (1999), 311–28.
7. Thomas Adams, *The Blacke Devil or the Apostate. Together with the Wolfe Worrying the Lambes and The Spirituall Navigator, Bound for the Holy Land* (London: 1615), pp. 31–2.
8. On Turner's career and affiliations, see Philip K. Wilson, *Surgery, Skin and Syphilis: Daniel Turner's London (1667–1741)* (Amsterdam; Atlanta: Rodopi, 1999).
9. Turner, *De Morbis Cutaneis*, p. 76.
10. See, for example: Nicholas Culpeper, *A Directory for Midwives, or, A Guide for Women, in their Conception, Bearing, and Suckling their Children* (London: 1651), pp. 165–6; Théophile Bonet, *A Guide to the Practical Physician* (London: 1684), p. 116.

11. Guy de Chauliac, *Grande Chirurgerie*, ed. E. Nicaise (Paris: 1890 [1363]), p. 305.
12. Turner, *De Morbis Cutaneis*, p. 75.
13. Luke Demaitre, 'Medieval Notions of Cancer: Malignancy and Metaphor', *Bulletin of the History of Medicine* 72:4 (1998), 616.
14. Richard Wiseman, *Several Chirurgical Treatises* (second edition) (London: 1686), p. 118.
15. Bruce Thomas Boehrer, *Animal Characters: Nonhuman Beings in Early Modern Literature* (Philadelphia: University of Pennsylvania Press, 2010), p. 165.
16. Edwards, 'Milton's Reformed Animals: An Early Modern Bestiary T–Z', 277–8.
17. Ibid.
18. Ibid.
19. Henry Smith, 'The Benefit of Contentation', in *The Sermons of Maister Henrie Smith* (London: 1593), p. 209.
20. Charles Cotton, 'Contentment: Pindarick Ode', in John Beresford (ed.), *Poems of Charles Cotton, 1630–1687* (London: Cobden-Sanderson, 1923), pp. 224–8.
21. Fudge, *Perceiving Animals*, pp. 51–5.
22. On indistinction between varieties of invertebrate in early modern texts, see MacInnes, 'The Politic Worm', especially p. 256.
23. Curiously, both the canker-worm as horticultural pest and the cancer-worm as disease agent are absent from Sujata Iyengar's entry on 'canker' in her *Shakespeare's Medical Language: A Dictionary* (London; New York: Continuum, 2011), pp. 51–4. This absence seems crucial to her more general downplaying of the 'ontology' of cancer, as discussed in Chapter 1.
24. Harris, 'The Canker of England's Commonwealth', 317.
25. Ibid.
26. Ibid., 317–18.
27. William Salmon, *Paraieremata, or Select Physical and Chirurgical Observations* (London: 1687), p. 378.
28. 'Sow' was used from the thirteenth century as a term for woodlice: see 'sow, n.1', *OED Online* http://www.oed.com, 27 March 2013.
29. D. Border, *Polypharmakos Kai Chymistes, or, The English Unparalell'd Physitian and Chyrurgian* (London: 1651), p. 15.
30. A.T., *A Rich Store-House or Treasury for the Diseased* (London: 1596), pp. 41–2; Mrs Corylon, *A Booke of Divers Medecines* (1606) Wellcome Library MS.213, p. 141r. See also: Elizabeth Sleigh and Felicia Whitfeld, *Collection of Medical Receipts* (1647–1722) Wellcome MS.751, p. 5; Sarah Hughes, *Mrs Hughes Her Receipts* (1637), Wellcome MS.363, p. 55r; Johanna St John, *Johanna St John Her Booke* (1680), Wellcome MS.4338, p. 14.
31. See Chapter 5.
32. Turner, *De Morbis Cutaneis*, p. 158.
33. Ibid.
34. Ibid.
35. On the advent and development of microscopy, see Catherine Wilson, *The Invisible World: Early Modern Philosophy and the Invention of the Microscope* (Princeton, NJ: Princeton University Press, 1997).

36. Pierre Dionis, *A Course of Chirurgical Operations, Demonstrated in the Royal Garden at Paris* (London: 1710 [French edition 1707]), p. 249.
37. MacInnes, 'The Politic Worm', p. 256. See also: Ambroise Paré, *The Case Reports and Autopsy Records of Ambroise Paré*, ed. Wallace B. Hamby (USA: Charles C. Thomas, 1960), which includes an account of one intestinal worm 'that resembled a serpent more than six feet long', yet was deemed 'not surprising' (p. 128); Daniel Le Clerc, *A Natural and Medicinal History of Worms, Bred in the Bodies of Men and Other Animals* (London: 1721). From *Open Library* (online resource), http://openlibrary.org/books, 26 April 2013.
38. Matthew Cobb, *The Egg and Sperm Race: The Seventeenth-Century Scientists Who Unravelled the Secrets of Sex, Life and Growth* (London: The Free Press, 2006), especially pp. 66, 84–9; MacInnes, 'The Politic Worm', especially pp. 255–6.
39. Thomas Moffet, *The Theater of Insects, or, Lesser Living Creatures*, appended to Edward Topsell, *The History of Four-Footed Beasts and Serpents* (London: 1658). Translated from *Insectorum, sive, Minimorum Animalium Theatrum* (London: 1634). On the provenance of Moffet's text, see Janice Neri, *The Insect and the Image: Visualizing Nature in Early Modern Europe, 1500–1700* (Minneapolis: University of Minnesota, 2011), pp. 45–65.
40. Moffet, *The Theater of Insects*, pp. 1100–6.
41. R. Clark, *Vermiculars Destroyed, with an Historical Account of Worms, Collected from the Best Authors as well Ancient as Modern, Proved by that Admirable Invention of the Microscope* (London: 1690). An advertisement for this text shows that it was first printed in 1661, though no extant copy remains. It was reprinted at least four times until 1691.
42. Ibid., pp. 11–14. See also William Ramesey, *Helminthologia, or, Some Physical Considerations of the Matter, Origination, and Several Species of Wormes Macerating and Direfully Cruciating Every Part of the Bodies of Mankind* (London: 1668).
43. Wilson, *The Invisible World*, pp. 70–80.
44. MacInnes, 'The Politic Worm', p. 263.
45. Interestingly, both the King James Bible (1611) and the Geneva Bible (1560) translate Joel 1:4 and 2:25 as featuring a 'cankerworm' which is absent from the same passages of the 1539 Great Bible. In turn, the King James Bible translates as 'cankerworm' in Nahum 3:15–16 the pest which appears as 'locust' in earlier versions including the Geneva, perhaps indicating a greater investment in that term as time wore on. A reference to cancer as a disease in 2 Timothy 2:17, however, remains stable throughout all three versions, as well as the 1526 Tyndale New Testament.
46. Harley, 'Last Things First in Chaucer's Physician's Tale', 6.
47. Ibid., 7. Quotation from a sermon by Richard Alkerton, c.1406.
48. See Wright, 'The World's Worst Worm', 121. For further examples of the worm as an agent of conscience, see also: Nicholas Billingsley, 'On Conscience', from *A Treasury of Divine Raptures* (1667), l.123–128; Henry Bold, 'Song XII', from *Latine Songs* (1685), p. 445; Benjamin Keach, 'Hymn 146: No Light, But Darkness There Doth Dwell', from *Spiritual Melody* (1691).
49. See, for example: Shakespeare's *Henry VI, Part 1* 3:1, in which the 'viperous worm / .. gnaws the bowels of the commonwealth'; John Milton, 'Arcades', in which the worm bites with 'cankered venom' (in *Milton: Complete Shorter Poems* (second edition), ed. John Carey (New York: Addison Wesley Longman, Inc., 1997), pp. 161–6, l.53).

50. Gordon Williams, *Shakespeare's Sexual Language: A Glossary,* 3 vols. (London; Atlantic Highlands, NJ: The Athlone Press, 1994), vol. 3, p. 1549.
51. This phrase is borrowed from Harris, '"The Canker of England's Commonwealth"', 318.
52. Gillian Bennet, 'Bosom Serpents and Alimentary Amphibians: A Language for Sickness', in Marijke Gijswit-Hofstra, Hilary Marland and Hans de Waardt (eds), *Illness and Healing Alternatives in Western Europe* (London; New York: Routledge, 1997), p. 225.
53. For an example, see the case of Rosemary Alvarez, whose 'brain worm' attracted international media coverage: 'Doctors Find Worm in Woman's Brain While Operating on "Tumour"', *Daily Mail,* 20 November 2008, http://www.dailymail.co.uk/news/article-1087937/Doctors-worm-womans-brain-operating-tumour.html; 'Live Worm "Burrowed through Woman's Brain"', *Nine MSN,* 21 November 2008, http://news.ninemsn.com.au/article. aspx?id=669745. Both accessed 22 January 2010.
54. Thomas R. Forbes, 'Verbal Charms in British Folk Medicine', *Proceedings of the American Philosophical Society* 115:4 (1971), 312.
55. Calvert Watkins, *How to Kill a Dragon: Aspects of Indo-European Poetics* (Oxford: Oxford University Press, 1995), especially chapters 57 and 58.
56. Ibid., p. 523.
57. Benjamin W. Fortson, *Indo-European Language and Culture: An Introduction* (second edition) (Singapore: Blackwell, 2011), p. 30.
58. James T. Patterson, *The Dread Disease: Cancer and Modern American Culture* (Cambridge, MA: Harvard University Press, 1987), p. 31. See also 'Fasting and Cancer: Starving the Beast', *The Economist* (9 February 2012). Accessed via www. economist.com/blogs/babbage/2012/02/fasting-and-cancer, 4 March 2013 (article); James Capuano, *Beast: A Slightly Irreverent Tale about Cancer (and Other Assorted Anecdotes)* (Wickford, RI: New Street Communications LLC, 2012); Cancer Research UK, 'Enemy' (advertisement) dir. Siri Bunford, 30 April 2013.

4　Cancerous Growth and Malignancy

1. 'malignant, adj. and n.'. *OED Online.* March 2013. Oxford University Press, http://www.oed.com/viewdictionaryentry/Entry/112926, 6 June 2013.
2. See for some examples: Kevin P. Siena, 'Pollution, Promiscuity, and the Pox: English Venereology and the Early Modern Discourse on Social and Sexual Danger', *Journal of the History of Sexuality* 8:4 (1998), 553–74; Louis F. Qualtiere and William W.E. Slights, 'Contagion and Blame in Early Modern England: The Case of the French Pox', *Literature and Medicine* 22:1 (2003), 1–24; Rebecca Totaro, *Suffering in Paradise: The Bubonic Plague in Literature from More to Milton* (Pittsburgh, Pennsylvania: Duquesne University Press, 2005); Margaret Healy, *Fictions of Disease in Early Modern England: Bodies, Plagues and Politics* (Basingstoke: Palgrave, 2001); Vivian Nutton, 'The Seeds of Disease: An Explanation of Contagion and Infection from the Greeks to the Renaissance', *Medical History* 27 (1983), 1–34.
3. Donald Beecher, 'An Afterword on Contagion', in Claire L. Carlin (ed.), *Imagining Contagion in Early Modern Europe* (Basingstoke: Palgrave Macmillan, 2005), p. 244.

4. Sarah Covington, *Wounds, Flesh and Metaphor in Seventeenth-Century England* (Basingstoke: Palgrave Macmillan, 2009); Colin Milburn, 'Syphilis in Faerie Land: Edmund Spenser and the Syphilography of Elizabethan England', *Criticism* 46:4 (2004), 597–632; David Harley, 'Medical Metaphors in English Moral Theology', *Journal of the History of Medicine and Allied Sciences* 48 (1993), 396–435.

5. See, for instance, Tanya Pollard, 'Enclosing the Body: Tudor Conceptions of Skin', in Kent Cartwright (ed.), *A Companion to Tudor Literature* (Oxford: Wiley-Blackwell, 2009), pp. 111–23; Lorraine Daston and Katherine Park, *Wonders and the Order of Nature, 1150–1750* (Cambridge, MA: Zone Books; distributed by MIT books, 1998), particularly Chapter 5, 'Monsters; A Case Study', pp. 173–214; David Cressy, 'Lamentable, Strange, and Wonderful: Headless Monsters in the English Revolution', in Laura Lunger Knoppers and Joan B. Landes (eds), *Monstrous Bodies/Political Monstrosities in Early Modern Europe* (Ithaca, NY: Cornell University Press, 2004), pp. 40–67.

6. Nicholas Culpeper, *A Directory for Midwives* (London: 1651), p. 324; John Pechey, Theodore Mayern (Sir Théodore Turquet de Mayerne), Dr. Chamberlain (probably Thomas Chamberlayne) and Nicholas Culpeper, *The Compleat Midwife's Practice* (London: 1698), p. 184.

7. Richard Wiseman, *Several Chirurgical Treatises* (second edition) (London: 1686), p. 101; William Beckett, *New Discoveries Relating to the Cure of Cancers.. To Which is Added, a Solution of Some Curious Problems, Concerning the Same Disease* (1711), pp. 19–20; *An Account of the Causes of Some Particular Rebellious Distempers viz. the Scurvey, Cancers in Women's Breasts, &c. Vapours, and Melancholy, &c. Weaknesses in Women, &c. Gout, Fistula in Ano, Dropsy, Agues, &c.* (London: 1670), p. 21; Ambroise Paré, *The Workes of that Famous Chirurgion Ambrose Parey* (trans. Thomas Johnson, book 29 trans. George Baker) (London: 1634), p. 279; Henri-François Le Dran, *Observations in Surgery* (trans. J[ohn] S[parrow]) (London: 1739), pp. 43–6.

8. *An Account of the Causes of Some Particular Rebellious Distempers*, pp. 26–7.

9. See Erwin H. Ackernecht, 'Historical Notes on Cancer', *Medical History* 2:2 (1958), 115–16.

10. Le Dran, *Observations in Surgery*, pp. 43–6.

11. Ibid. See also Gendron, *Enquiries*, p. 82; Théophile Bonet, *A Guide to the Practical Physician* (London: 1684), p. 63. Wiseman's *Several Chirurgical Treatises* also records a case of cancer recurring after seven years, subsequently killing the patient (p. 115).

12. Peter Lowe, *The Whole Course of Chirurgerie* (London: 1597), sig. Aa1r.

13. Pechey et al., *The Compleat Midwife's Practice*, p. 185; Pierre Dionis, *A Course of Chirurgical Operations, Demonstrated in the Royal Garden at Paris* (London: 1710 (French edition 1707]), p. 248.

14. The other 'lips' to cast out noxious matter (vaginal discharge or menstrual blood) were those of the female genitalia – on the gendering of cancer and its relation to the womb, see Chapter 2.

15. Dionis, *A Course of Chirurgical Operations*, p. 250. See also Pechey et al., *The Compleat Midwife's Practice*, pp. 184–5; Culpeper, *A Directory for Midwives*, p. 324; Alexander Read, *The Chirurgicall Lectures of Tumors and Ulcers* (London: 1635), pp. 214–15; Charles Gabriel Le Clerc, *A Description of Bandages and*

Dressings, According to the Most Commodious Ways Now Used in France (London: 1701), pp. 55–7.

16. Lowe, *The Whole Course of Chirurgerie*, sig. L3r–L4r on cancerous tumours, sig. Aa1r–Aa1v on cancerous ulcers; John Pechey, *The Store-House of Physical Practice* (London: 1695), p. 116. See also Luke Demaitre, 'Medieval Notions of Cancer: Malignancy and Metaphor', *Bulletin of the History of Medicine* 72:4 (1998), 612, which establishes the ulcerate/non-ulcerate distinction as one also employed in medieval texts.

17. Henry More, 'Letter from Henry More to Lady Anne Conway, 31 December 1674', in Anne Conway, Henry More et al., *The Conway Letters: The Correspondence of Anne, Viscountess Conway, Henry More, and their Friends, 1642–1684*, ed. Marjorie Hope Nicolson. Revised by Sarah Hutton (New York: Oxford University Press, 1992 [1930]), pp. 398–9. The medicine to which Dr. Clark refers is one sent to the sick woman's household by François-Mercure Van Helmont, son of Jean-Baptiste Van Helmont and a correspondent of both More and Conway.

18. Wiseman, *Several Chirurgical Treatises*, p. 101; Jane Sharp, *The Midwives Book* (London: 1671), pp. 346–7. See also Culpeper, *A Directory for Midwives*, pp. 324–6; Gendron, *Enquiries*, p. 19, 21.

19. Notably, the words 'venom' or 'poison' in these contexts operated as both descriptive and categorical terms. For *venom* (n.), the *Oxford English Dictionary* lists '1. The poisonous fluid normally secreted by certain snakes and other animals and used by them in attacking other living creatures.. 2. Poison, especially as administered to or drunk by a person; any poisonous or noxious substance, preparation, or property; a morbid secretion or virus. Now rare. 3. a. *fig.* Something comparable to or having the effect of poison; any baneful, malign, or noxious influence or quality; bitter or virulent feeling, language, etc.' *Oxford English Dictionary Online*, http://oed.com/viewdictionaryentry/Entry/222182, 19 July 2012. On the use of 'infection' to mean 'poisoned' or 'envenomed', see Roger Lund, 'Infectious Wit: Metaphor, Atheism and the Plague in Eighteenth-Century London', *Literature and Medicine* 22:1 (2003), 45–64.

20. Pechey et al., *The Compleat Midwife's Practice*, p. 183. See also Johann Jacob Wecker, *A Compendious Chyrurgerie: Gathered, & Translated (especially) out of Wecker* (trans. with additions by John Banister) (London: 1585), p. 106; Wiseman, *Several Chirurgical Treatises*, pp. 98–9.

21. Jacques Guillemeau, 'A.H.', and W. Bailey, *A Worthy Treatise of the Eyes.. Togeather With a Profitable Treatise of the Scorbie; & Another of the Cancer by A. H.* (London: 1587), pp. 61–2.

22. Browne, *The Surgeons Assistant*, pp. 104–5.

23. Robert Boyle, *Some Considerations Touching the Usefulnesse of Experimental Naturall Philosophy Propos'd in Familiar Discourses to a Friend, By Way of Invitation to the Study of it* (Oxford: 1663), pp. 57–8 (author's italics).

24. *An Account of the Damnable Prizes in Old Nicks Lottery, for Men of Honour Only; Where Every Man that Ventures, is Sure to Get the Lord Knows What For Ever. In a Gradation of Familiar Thoughts, Arising, Upon the Not Passing of the Duelling Bill, Brought in Last Session of Parliament* (London: 1712), p. 3.

25. *An Account of the Causes of Some Particular Rebellious Distempers*, p. 23.

26. Beckett, *New Discoveries*, p. 11.

27. Miranda Wilson, 'Watching Flesh: Poison and the Fantasy of Temporal Control in Renaissance England', *Renaissance Studies* 27:1 (2013), 97–113.
28. *An Account of the Causes of Some Particular Rebellious Distempers*, pp. 24–5.
29. Marjo Kaartinen, *Breast Cancer in the Eighteenth Century* (London; Vermont: Pickering and Chatto, 2013), p. 20.
30. Beckett, *New Discoveries*, p. 36.
31. Ibid., p. 37.
32. See Marie E. McAllister, 'Stories of the Origin of Syphilis in Eighteenth-Century England: Science, Myth, and Prejudice', *Eighteenth-Century Life* 24:1 (2000), 22.
33. See Kevin P. Siena, *Venereal Disease, Hospitals and the Urban Poor: London's 'Foul Wards', 1600–1800* (Rochester, NY: University of Rochester Press, 2004), pp. 65–6.
34. Browne, *The Surgeons Assistant*, p. 78.
35. Luke Demaitre, *Leprosy in Premodern Medicine: A Malady of the Whole Body* (Baltimore: Johns Hopkins University Press, 2007), p. 191.
36. Demaitre, 'Medieval Notions of Cancer', particularly pp. 609–12; Philip Barrough, *The Method of Physick* (London: 1583), p. 275.
37. From Ambroise Paré's *Le Oeuvres* (1575) cited in Demaitre, *Leprosy in Premodern Medicine*, p. 249.
38. Gideon Harvey, *Little Venus Unmask'd* (seventh edition) (London: 1702 [first edition 1670]), p. 65. See also Salvator Winter, *A New Dispensary of Fourty Physicall Receipts, Most Necessary and Profitable for all House-Keepers in their Families* (London: 1649), pp. 13–14.
39. Harry Keil, 'The Evolution of the Term Chancre and its Relation to the History of Syphilis', *Journal of the History of Medicine* (1949), 407–9.
40. Siena, *Venereal Disease, Hospitals and the Urban Poor*, particularly pp. 22–5.
41. Everard Maynwaringe, *The Frequent, but Unsuspected Progress of Pains, Inflammations, Tumors, Apostems, Ulcers, Cancers, Gangrenes and Mortifications* (London: 1679), pp. 194–5; Dionis, *A Course of Chirurgical Operations*, pp. 256–7.
42. Qualtiere and Slights, 'Contagion and Blame in Early Modern England', 6.
43. Browne, *The Surgeons Assistant*, p. 86.
44. Ibid., pp. 111–12.
45. Paré, *The Workes*, p. 281. See also Jean Riolan, *A Sure Guide, or, the Best and Nearest Way to Physick and Chyrurgery* (trans. Nicholas Culpeper and W.R.) (London: 1657), p. 97; Sharp, *The Midwives Book*, pp. 346–7.
46. See Vivian Nutton, 'The Seeds of Disease: An Explanation of Contagion and Infection from the Greeks to the Renaissance', *Medical History* 27 (1983), 1–34; Lucinda Cole, 'Of Mice and Moisture: Rats, Witches, Miasma, and Early Modern Theories of Contagion', *Journal for Early Modern Cultural Studies* 10:2 (2010), 65–84.
47. Beckett, *New Discoveries*, pp. 39–41.
48. Ibid.
49. Ibid.
50. See Siena, 'Pollution, Promiscuity, and the Pox', 561; Siena, *Venereal Disease*, p. 17, 193; also Jacques Guillemeau, *Childbirth, or, the Happy Delivery of Women..To Which is Added, a Treatise of the Diseases of Infants, and Young*

Children: With the Cure of Them (London: 1612 [first edition in French c.1609]), sig. Kk1v.

51. Beckett, *New Discoveries*, pp. 41–3.
52. Ibid.
53. Kaartinen, *Breast Cancer in the Eighteenth Century*, p. 19.
54. Barrough, *The Method of Physick*, p. 274.
55. Ibid.
56. Karen Edwards, 'Milton's Reformed Animals: An Early Modern Bestiary A-C', *Milton Quarterly* 39:4 (2005), 249.
57. Shakespeare, 'Sonnet 35'. On this subject, see my chapter 'The Worm and the Flesh: Cankered Bodies in Shakespeare's Sonnets', in Sujata Iyengar (ed.), *Disability, Health and Happiness in the Shakespearean Body* (New York; Abingdon, UK: Routledge, 2015), pp. 240–60.
58. On 'discovering' cancers, see: *An Account of the Causes of Some Particular Rebellious Distempers*, p. 21; Le Dran, *Observations in Surgery*, p. 44; Gendron, *Enquiries*, pp. 12–13; Daniel Turner, *De Morbis Cutaneis: Diseases Incident to the Skin* (London: 1714), p. 76.
59. Gendron, *Enquiries*, pp. 12–13, 16–17.
60. See the *Oxford English Dictionary Online* entry: 'discover, v.'. OED Online, June 2012, Oxford University Press, 22 August 2012, http://oed.com/view/ Entry/54005, 15 October 2012.
61. Dionis, *A Course of Chirurgical Operations*, pp. 247–8.
62. *An Account of the Causes of Some Particular Rebellious Distempers*, p. 21.
63. Bonet, *A Guide to the Practical Physician*, p. 62.
64. *An Account of the Causes of Some Particular Rebellious Distempers*, p. 20; John Browne, *Adenochoiradelogia, or, an Anatomick-Chirurgical Treatise of Glandules & Strumaes or, Kings-Evil-Swellings* (London: 1684), pp. 31–2; Barrough, *The Method of Physick*, p. 273. See also Wiseman, *Several Chirurgical Treatises*, pp. 98–9; Paré, *The Workes*, p. 280; John Tanner, *The Hidden Treasures of the Art of Physic* (1659), p. 443; Turner, *De Morbis Cutaneis*, p. 75.
65. See Thomas D'Urfey, *Madam Fickle: Or the Witty False One. A Comedy* (London: 1677), 5.1.
66. Wither, 'Opobalsamum Anglicanum'.
67. Gerrard Malynes, *A Treatise of the Canker of Englands Common-Wealth* (London: 1601), pp. 18, 19.
68. John Fletcher, *The Faithful Shepherdess* (1679 [1608]), 5.3. From *English Drama* (online resource), www.0-collections.chadwyck.co.uk, 30 September 2012.
69. Harley, 'Medical Metaphors', 408 [quoting from Daniel Getsius, *Tears Shed in the Behalf of his Dear Mother the Church of England, and Her Sad Distractions: Gathered and Brought Into this Small Paper Vessell for the Use of the Vulgar, and Common People, Not to Play with Religion* (1658), pp. 116–17].
70. Thomas Adams, *The Blacke Devil or the Apostate. Together with the Wolfe Worrying the Lambes and The Spirituall Navigator, Bound for the Holy Land* (London: 1615), p. 31.
71. *An Account of the Damnable Prizes in Old Nick's Lottery*, p. 3.
72. Andy Wood, *Riot, Rebellion and Popular Politics in Early Modern England* (Basingstoke: Palgrave, 2002), p. 27.
73. Camilla Rockwood, (ed.), 'Malignants: A Term Applied by the Parliamentarians to the Royalists Who Fought for Charles I and Charles II in the Civil Wars',

Brewer's Dictionary of Phrase & Fable, eighteenth edition (London: Chambers Harrap Publishers Ltd., 2010), p. 829.

74. See Kaartinen, *Breast Cancer in the Eighteenth Century*, especially pp. 7–8; Daniel De Moulin, 'Historical Notes on Breast Cancer, With Emphasis on the Netherlands: II. Pathophysiological Concepts, Diagnosis and Therapy in the 18th Century', *The Netherlands Journal of Surgery* 33:4 (1981), 206–16.

5 Wolves' Tongues and Mercury: Pharmaceutical Cures for Cancer

1. Richard Wiseman, *Several Chirurgical Treatises* (second edition) (London: 1686), p. 102.
2. On 'Oyl of Frogs', see ibid., p. 102.
3. Marjo Kaartinen, *Breast Cancer in the Eighteenth Century* (London; Vermont: Pickering and Chatto, 2013), especially pp. 27–35.
4. Ibid., pp. 28, 35.
5. Ibid., pp. 58–60.
6. Ibid., pp. 30–1.
7. Ibid., pp. 31–2.
8. Wendy D. Churchill, *Female Patients in Early Modern Britain: Gender, Diagnosis, and Treatment* (Farnham: Ashgate, 2012), pp. 128–30. See also: Michael Stolberg, *Experiencing Illness and the Sick Body in Early Modern Europe* (Basingstoke: Palgrave Macmillan, 2011), pp. 135–9; Alasdair B. MacGregor, 'The Search for a Chemical Cure for Cancer', *Medical History* 10 (1966), 375.
9. Luke Demaitre, 'Medieval Notions of Cancer: Malignancy and Metaphor', *Bulletin of the History of Medicine* 72:4 (1998), 631. On the *Dreckapotheke*, see Markham Judah Geller, *Volume 7: Renal and Rectal Disease Texts* from the series *Die Babylonisch-Assyrische Medizin in Texten und Untersuchungen* (ed. Robert Biggs) (Germany: De Gruyter, 2005), p. 7.
10. Siddhartha Mukherjee, *The Emperor of All Maladies: A Biography of Cancer* (London: Scribner, 2010), p. 50.
11. See Kevin P. Siena, *Venereal Disease, Hospitals and the Urban Poor: London's 'Foul Wards', 1600–1800* (Rochester, NY: University of Rochester Press, 2004), especially pp. 22–7; Jon Arrizabalaga, John Henderson and Roger French, *The Great Pox: The French Disease in Renaissance Europe* (New Haven; London: Yale University Press, 1997), especially pp. 139–88.
12. Peter Elmer, 'Chemical Medicine and the Challenge to Galenism: The Legacy of Paracelsus, 1560–1700', in Peter Elmer (ed.), *The Healing Arts: Health, Disease and Society in Europe, 1500–1800* (Manchester: Manchester University Press, 2004), pp. 108–35; Andrew Wear, *Knowledge and Practice in English Medicine, 1550–1680* (Cambridge: Cambridge University Press, 2000), especially Chapter 10, 'Changes and Continuities', pp. 434–73.
13. Wear, *Knowledge and Practice*, p. 23.
14. John Fernelius, *Select Medicinal Counsels of John Fernelius, Chief Physitian to the King*. Appended to Felix Platter, Abdiah Cole and Nicholas Culpeper, *A Golden Practice of Physick* (London: 1662), pp. 412–13; image 742.
15. Ibid.

16. Wear, *Knowledge and Practice*, especially Chapter 4, 'Preventive Medicine: Healthy Lifestyles and Healthy Environments', pp. 154–209.
17. Jan Purnis, 'The Stomach and Early Modern Emotion', *University of Toronto Quarterly* 79:2 (2010), 807. See also Michael Schoenfeldt, *Bodies and Selves in Early Modern England: Physiology and Inwardness in Spenser, Shakespeare, Herbert, and Milton* (Cambridge: Cambridge University Press, 1999), especially pp. 20–30.
18. Margaret Healy, 'Bodily Regimen and Fear of the Beast: "Plausibility" in Renaissance Domestic Tragedy', in Erica Fudge, Ruth Gilbert and Susan Wiseman (eds), *At the Borders of the Human: Beasts, Bodies and Natural Philosophy in the Early Modern Period* (Basingstoke: Palgrave Macmillan, 1999), pp. 58–9.
19. Ken Albala, *Eating Right in the Renaissance* (Berkeley, CA: University of California Press, 2002), in particular Chapter 8, 'Medicine and Cuisine', pp. 241–83. See also Joan Thirsk, *Food in Early Modern England: Phases, Fads, Fashions, 1500–1760* (London: Continuum, 2009).
20. Galen of Pergamon, *Certaine Workes of Galens, Called Methodus Medendi* (trans. Thomas Gale) (London: 1566), pp. 54–7.
21. Philip Barrough, *The Method of Physick* (London: 1583), p. 275.
22. Ibid.
23. Galen, *Methodus Medendi*, p. 55.
24. John Pechey, Theodore Mayern (Sir Théodore Turquet de Mayerne), Dr. Chamberlain (probably Thomas Chamberlayne) and Nicholas Culpeper, *The Compleat Midwife's Practice* (London: 1698), p. 165. See also: Johann Jacob Wecker, *A Compendious Chyrurgerie: Gathered, & Translated (especially) out of Wecker* (trans. with additions by John Banister) (London: 1585), p. 108; Wiseman, *Several Chirurgical Treatises*, p. 99.
25. Ambroise Paré, *The Workes of that Famous Chirurgion Ambrose Parey* (trans. Thomas Johnson, book 29 trans. George Baker) (London: 1634), pp. 281–2.
26. Alexander Read, *The Chirurgicall Lectures of Tumors and Ulcers* (London: 1635), p. 212 (pagination is irregular). On the dangers of strong wines, see also: Alexander Read, *The Workes of that Famous Physician Dr. Alexander Read* (second edition) (London: 1650), p. 248.
27. Barrough, *The Method of Physick*, p. 173.
28. Galen, *Methodus Medendi*, p. 54. On the dangers of strenuous exercise, see: John Sadler, *The Sicke Woman's Private Looking-Glasse* (London: 1636), p. 29; Peter Lowe, *The Whole Course of Chirurgerie* (London: 1597), sig. L3v; Christof Wirsung, *Praxis Medicinae Universalis* (trans. Jacob Mosan) (London: 1598), p. 572; Read, *The Chirurgicall Lectures*, p. 215.
29. Elizabeth A. Williams, 'Sciences of Appetite in the Enlightenment, 1750–1800', *Studies in History and Philosophy of Biological and Biomedical Sciences* 43 (2012), 402.
30. See, for example, Read, *The Workes*, pp. 172–3.
31. Sarah Toulalan, '"To[o] Much Eating Stifles the Child": Fat Bodies and Reproduction in Early Modern England', *Historical Research* 87:235 (2014), 67–93. See also: Gail Kern Paster, *The Body Embarrassed: Drama and the Disciplines of Shame in Early Modern England* (Ithaca, NY: Cornell University Press, 1993), especially pp. 65–85.
32. Healy, 'Bodily Regimen and Fear of the Beast', pp. 57–8.

33. Galen, *Methodus Medendi*, pp. 55–6. This was directly opposite to the advice for melancholy individuals, underlining the perceived importance of heating or 'burning' in the creation of *atra bilis*.
34. Read, *The Workes*, p. 172.
35. William Beckett, *New Discoveries Relating to the Cure of Cancers .. To Which is Added, a Solution of Some Curious Problems, Concerning the Same Disease* (1711), p. 25.
36. Ibid.
37. Lazarus Riverius, *Four Books of that Learned and Renowned Doctor, Lazarus Riverius*. Appended to Felix Platter, Abdiah Cole and Nicholas Culpeper, *A Golden Practice of Physick* (London: 1662), pp. 55–6; image 610–11. See the Bibliography for notes on this composite text.
38. Lazarus Riverius, *The Practice of Physick* (trans. with possible additions by Nicholas Culpeper, Abdiah Cole and William Rowland) (London: 1655), p. 88.
39. See, for example: John Pechey, *The Store-House of Physical Practice* (1695), p. 62; Nicholas Fontanus, *The Womans Doctour* (London: 1652), pp. 114–16; Joannes Jonstonus, *The Idea of Practicall Physick* (1657), p. 8.
40. See also Stolberg, who views purging as a practice related to exorcism (*Experiencing Illness*, p. 27).
41. Riverius, *The Practice of Physick*, p. 492; John Browne, *The Surgeons Assistant .. Also a Compleat Treatise of Cancers and Gangreens. With an Enquiry Whether they have any Alliance with Contagious Diseases* (London: 1703), pp. 89–90.
42. Paster, *The Body Embarassed*, especially Chapter 2, 'Laudable Blood: Bleeding, Difference, and Humoral Embarrasment', pp. 64–112.
43. On phlebotomy bringing about miscarriage, see Cathy McClive, 'The Hidden Truths of the Belly: The Uncertainties of Pregnancy in Early Modern Europe', *Social History of Medicine* 15:2 (2002), 224–5.
44. Browne, *The Surgeons Assistant*, pp. 89–90.
45. Paul Barbette with Raymundus Minderius (Raymond Minderer), *Thesaurus Chirurgiae* (fourth edition) (London: 1687 [1676]), p. 124; Wirsung, *Praxis Medicinae Universalis*, p. 527.
46. Paster, *The Body Embarassed*, pp. 83–4.
47. Schoenfeldt, *Bodies and Selves in Early Modern England*, pp. 31–3.
48. See in particular Eve Keller, '"That Sublimest Juyce in our Body": Bloodletting and Ideas of the Individual in Early Modern England', *Philological Quarterly* 86:1/2 (2007), 97–123.
49. Ibid.
50. Fernelius, *Select Medical Counsels*, pp. 412–13; image 742.
51. Barrough, *The Method of Physick*, pp. 275–6.
52. Ibid., p. 276.
53. Nicholas Culpeper, *A Directory for Midwives* (London: 1651), p. 324. See also: Paré, *The Workes*, p. 281; Robert Bayfield, *Tractatus de Tumoribus Praeter Naturam, or, A Treatise of Preternatural Tumors* (London: 1662), p. 187; Jacques Guillemeau, 'A.H.', and W. Bailey, *A Worthy Treatise of the Eyes .. Togeather with a Profitable Treatise of the Scorbie; & Another of the Cancer by A. H.* (London: 1587), p. 82; Barbette, *Thesaurus Chirurgiae*, p. 123.
54. Wirsung, *Praxis Medicinae Universalis*, p. 98.

55. See, for examples: Browne, *The Surgeons Assistant*, pp. 94–5; Barrough, *The Method of Physick*, p. 276; Barbette, *Thesaurus Chirurgiae*, p. 124; John Smith, *A Compleat Practice of Physick* (1656), pp. 51–2; Pechey et al., *The Compleat Midwife's Practice*, p. 185; Daniel Sennert, Nicholas Culpeper and Abdiah Cole, *Practical Physick: The Fourth Book, in Three Parts* (London: 1664), p. 215.

56. Nicholas Culpeper, *The English Physitian Enlarged* (London: 1653 [first published 1652]), pp. 300–1.

57. Ibid., pp. 315–19.

58. Ibid., pp. 124–5 (Henbane), 172–3 (Nightshade).

59. Elizabeth Godfrey, *Collection of Medical and Cookery Receipts* (1686), Wellcome MS.2535, pp. 106–7.

60. Pechey et al., *The Compleat Midwife's Practice*, pp. 220–1; Barbette, *Thesaurus Chirurgiae*, p. 124; Paré, *The Workes*, p. 283. See also: *A Physical Dictionary* (London: 1657), p. 68; Browne, *The Surgeons Assistant*, pp. 93–4, 97; Smith, *A Compleat Practice of Physic*, pp. 50–1; Guillemeau, 'A.H.' and Bailey, *A Worthy Treatise*, p. 44.

61. Michael B. Shimkin, *Contrary to Nature: Being an Illustrated Commentary on Some Persons and Events of Historical Importance in the Development of Knowledge Concerning Cancer* (Washington, 1977), p. 32; A. Kaprozilos and N. Pavlidis, 'The Treatment of Cancer in Greek Antiquity', *European Journal of Cancer* 40 (2004), 2033–40.

62. This practice continued into the eighteenth century: see Kaartinen, *Breast Cancer in the Eighteenth Century*, p. 25.

63. Oswald Gabelkover, *The Boock of Physicke* (trans. 'A.M.') (Dorte [probably modern Dortmund]: 1599), p. 367.

64. D. Border, *Polypharmakos Kai Chymistes, or, the English Unparalell'd Physitian and Chyrurgian* (London: 1651), p. 15.

65. Harold J. Cook, *The Decline of the Old Medical Regime in Stuart London* (New York: Cornell University Press, 1986), p. 68.

66. Daniel Turner, *De Morbis Cutaneis: Diseases Incident to the Skin* (London: 1714), p. 76.

67. Ibid.

68. M. Robinson, 'For the Publick Good' (advertisement), in *The Original Weekly Journal: With Fresh Advices, Foreign and Domestick* (11–18 May 1717), p. 806. See also *The St. James's Evening Post* 288 (12 March 1717), p. 9. Both from *Eighteenth Century Journals*, www.amdigital.co.uk, 28 September 2012.

69. Katherine Jones (Viscountess Ranelagh), *Collection of Medical Receipts, c. 1675–c.1710*, Wellcome Library MS.1340, p. 128; Johanna St John, *Johanna St. John Her Booke* (1680), Wellcome Library MS.4338, pp. 14, 218.

70. Barrough, *The Method of Physick*, p. 276.

71. William Clowes, *A Short and Profitable Treatise Touching the Cure of the Disease Called Morbus Gallicus by Unctions* (1579), sig. D3v.

72. Giovannida Vigo, *The Most Excellent Workes of Chirurgerie* (London: 1571 [1543]), pp. 267–7.

73. Ibid.

74. Paul Dubé, *The Poor Man's Physician and Surgeon* (London: 1704), p. 362; Culpeper, *A Directory for Midwives*, pp. 165–6.

75. Barrough, *The Method of Physick*, p. 275; Pechey et al., *The Compleat Midwife's Practice*, p. 183; John Tanner, *The Hidden Treasures of the Art of Physic* (1659), p. 443.

76. Paré, *The Workes*, pp. 61–2.
77. George Hartman, *The Family Physitian* (London: 1696), pp. 245–6.
78. See Elizabeth Grey, Countess of Kent, *A Choice Manual of Rare and Select Secrets* (London: 1653), p. 34; Alethea Talbot, Countess of Arundel, *Natura Exenterata: Or Nature Unbowelled by the Most Exquisite Anatomizers of her* (London: 1655), p. 20.
79. 'chemotherapy, n.'. *OED* Online, December 2012, http://www.oed.com, 7 May 2013.
80. Grey, *A Choice Manual*, p. 51. See also: Read, *The Chirurgicall Lectures*, p. 218; William Salmon, *Paraieremata, or Select Physical and Chirurgical Observations* (London: 1687), pp. 277–8: Smith, *A Compleat Practice of Physic*, pp. 50–1.
81. Siena, *Venereal Disease*, pp. 22–6.
82. Théophile Bonet, *A Guide to the Practical Physician* (London: 1684), p. 62.
83. Vigo, *The Most Excellent Works*, p. xlvi.
84. Theodoric Borgognoni, *The Surgery of Theodoric* (*c.*1267) (trans. Eldridge Campbell) (New York: Appleton-Century-Crofts, Inc., 1955), quoted in Shimkin, *Contrary to Nature*, p. 42.
85. Culpeper, *A Directory for Midwives*, pp. 325–6; Riverius, *Four Books*, p. 299; Read, *The Workes*, p. 117.
86. Ruth Kleinmann, 'Facing Cancer in the Seventeenth Century: The Last Illness of Anne of Austria, 1644–1666', *Advances in Thanatology* 4 (1978), 43.
87. Read, *The Workes*, p. 174; Lowe, *The Whole Course of Chirurgerie*, sig. Aa1v.
88. See Sujata Iyengar, *Shakespeare's Medical Language: A Dictionary* (London; New York: Bloomsbury, 2011), pp. 286–7.
89. Siena, *Venereal Disease*, p. 23.
90. Ibid., p. 23.
91. Bonet, *A Guide to the Practical Physician*, p. 62.
92. See Bayfield, *Tractatus de Tumoribus*, p. 189.
93. William Beckett, *New Discoveries Relating to the Cure of Cancers..to which is added, a Solution of Some Curious Problems, Concerning the Same Disease* (second edition) (1712), pp. 53–4.
94. Ibid., p. 57.
95. Browne, *The Surgeons Assistant*, pp. 102–4. See also Wiseman, *Several Chirurgical Treatises*, p. 10.
96. Bonet, *A Guide to the Practical Physician*, p. 62.
97. Ibid., p. 62.
98. On the judgement and punishment of malpractice, see Harold Cook, *Trials of an Ordinary Doctor: Joannes Groenevelt in Seventeenth-Century London* (Baltimore: Johns Hopkins University Press, 1994), especially pp. 1–25.
99. Ward, *Diary*, p. 250.
100. Michael Schoenfeldt, 'Aesthetics and Anaesthetics: The Art of Pain Management in Early Modern England', in Jan Frans van Dijkhuizen and Karl A.E. Enenkel (eds), *The Sense of Suffering; Constructions of Physical Pain in Early Modern Culture* (Leiden; Boston: Brill, 2009), pp. 19–38.
101. David N. Harley, 'Medical Metaphors in English Moral Theology', *Journal of the History of Medicine and Allied Sciences* 48 (1993), 396–435. See also David Harley, 'Spiritual Physic, Providence and English Medicine, 1560–1640', in Peter Ole Grell and Andrew Cunningham (eds), *Medicine and the Reformation* (Abingdon: Routledge, 1993), pp. 101–17; and Andrew Wear, 'Puritan Perceptions of Illness in Seventeenth-Century England', in Roy Porter (ed.),

Patients and Practitioners: Lay Perceptions of Medicine in Pre-Industrial Society (Cambridge: Cambridge University Press, 1985), pp. 54–99.

102. Porter, *Bodies Politic*, p. 116.
103. Browne, *The Surgeons Assistant*, pp. 102–4.
104. Wiseman, *Several Chirurgical Treatises*, p. 10.
105. Henry More, 'Letter from Henry More to Lady Conway, Sept 17, 1674', in Anne Conway, Henry More et al., *The Conway Letters: The Correspondence of Anne, Viscountess Conway, Henry More, and their Friends, 1642–1684*, ed. Marjorie Hope Nicolson. Revised by Sarah Hutton (New York: Oxford University Press, 1992 [1930]), p. 392.
106. Henry More, 'Letter from Henry More to Lady Conway, October 19, 1674', in Conway, More et al., *The Conway Letters*, pp. 393–4.
107. Henry More, 'Letter from Henry More to Lady Conway, December 9, 1674', in Conway, More et al., *The Conway Letters*, p. 398.
108. Henry More, 'Letter from Henry More to Lady Conway, December, 31, 1674', in Conway, More et al., *The Conway Letters*, pp. 398–9.
109. Henry More, 'Letter from Henry More to Lady Conway, March 22, 1676', in Conway, More et al., *The Conway Letters*, p. 426.

6   'Cannot You Use a Loving Violence?': Cancer Surgery

Reverend John Ward, *Diary of the Rev. John Ward, A.M., Extending from 1648 to 1679*, ed. Charles Severn (London: Henry Coldurn, 1839), p. 250. From *Internet Archive* (online resource), http://www.archive.org, 2 March 2012.

1. Jacobus de Voragine, *The Golden Legend: Selections* (selected and trans. Christopher Stace, with an Introduction and Notes by Richard Hamer) (Harmondsworth: Penguin, 1998), p. 78.
2. Ibid., p. 77.
3. Edward F. Lewison, 'Saint Agatha, the Patron Saint of Diseases of the Breast, in Legend and Art', *Bulletin of the History of Medicine* 24 (1950), 409–20.
4. Richard Hamer, 'Introduction' in Voragine, *The Golden Legend*, p. ix.
5. Liana de Girolama Cheney, 'The Cult of St. Agatha', *Women's Art Journal* 17:1 (1996), 4–5.
6. See George T. Pack, 'St Peregrine, O.S.M. – The Patron Saint of Cancer Patients', *CA: A Cancer Journal for Clinicians* 17 (1967), 183–4.
7. Richard Wiseman, *Several Chirurgical Treatises* (second edition) (London: 1686), p. 113.
8. Siddhartha Mukherjee, *The Emperor of All Maladies: A Biography of Cancer* (London: Scribner, 2010), pp. 39–41; James S. Olson, *Bathsheba's Breast: Women, Cancer and History* (Baltimore, MD: Johns Hopkins University Press, 2002), p. 10; William L. Donegan, 'An Introduction to the History of Breast Cancer', in William L. Donegan and John Stricklin Spratt (eds), *Cancer of the Breast* (Philadelphia; London: Elsevier Science, 2002), p. 2; Michael B. Shimkin, *Contrary to Nature: Being an Illustrated Commentary on some Persons and Events of Historical Importance in the Development of Knowledge Concerning Cancer* (Washington: 1977), p. 22. See also Harold Ellis, *The Cambridge Illustrated History of Surgery* (Cambridge; New York: Cambridge University Press, 2009), pp. 165–9, which provides a basic timeline of the long-term development of cancer surgery.

9. Luke Demaitre, 'Medieval Notions of Cancer: Malignancy and Metaphor', *Bulletin of the History of Medicine* 72:4 (1998), 631–2.

10. Marie-Christine Pouchelle, *The Body and Surgery in the Middle Ages* (trans. Rosemary Morris) (Cambridge: Polity Press, 1990), p. 72.

11. Marjo Kaartinen, *Breast Cancer in the Eighteenth Century* (London; Vermont: Pickering and Chatto, 2013), pp. 41–54. See also Wendy D. Churchill, *Female Patients in Early Modern Britain: Gender, Diagnosis, and Treatment* (Farnham: Ashgate, 2012), pp. 130–8.

12. Roy Porter and Dorothy Porter, *In Sickness and in Health: The British Experience 1650–1850* (London: Fourth Estate, 1988), p. 106.

13. Lynda Ellen Stephenson Payne, *With Words and Knives: Learning Medical Dispassion in Early Modern England* (Aldershot: Ashgate, 2007).

14. Andrew Wear, *Knowledge and Practice in English Medicine, 1550–1680* (Cambridge: Cambridge University Press, 2000), in particular Chapter 5, 'Surgery: The Hand Work of Medicine', pp. 210–74; Philip K. Wilson, *Surgery, Skin and Syphilis Daniel Turner's London (1677–1741)* (Amsterdam: Rodopi, 1999).

15. Wear, *Knowledge and Practice*, p. 220.

16. Ibid., p. 249.

17. Wilson, *Surgery, Skin and Syphilis*, p. 94.

18. See in particular Jonathan Sawday, *The Body Emblazoned: Dissection and the Human Body in Renaissance Culture* (London: Routledge, 1995), especially pp. 45–50; Roger French, *Dissection and Vivisection in the European Renaissance* (Aldershot; Vermont: Ashgate, 1999), especially pp. 2–7; Florike Egmond, 'Execution, Dissection, Pain and Infamy – A Morphological Investigation', in Florike Egmond and Robert Zwijnenberg (eds), *Bodily Extremities: Preoccupations with the Human Body in Early Modern European Culture* (Aldershot: Ashgate, 2003), pp. 92–126; Richard Sugg, *Murder after Death: Literature and Anatomy in Early Modern England* (Ithaca, NY: Cornell University Press, 2007).

19. James Handley, *Colloquia Chirurgica: Or, the Whole Art of Surgery Epitomiz'd and made Easie, According to Modern Practice* (London: 1705), p. 70.

20. Kaartinen, *Breast Cancer in the Eighteenth Century*, p. 39.

21. Samuel Pepys, *Diary* (5 May 1665, n.p). From *The Diary of Samuel Pepys* (online resource), http://www.pepysdiary.com, 5 February 2013.

22. *London Evening Post*, Issue 27 (London: 8–10 February 1728). See also *London Evening Post*, Issue 135 (London: 17–19 October 1728); *Post Boy*, Issue 4008 (London: 7–9 April 1715). All from *Burney Newspaper Collection* (online resource), http://.galegroup.com, 4 March 2013.

23. See, for example: *London Evening Post*, Issue 187 (15–18 February 1729); *Daily Journal*, Issue 5828 (28 October 1736); *Daily Gazetteer* (London Edition), Issue 423 (3 November 1736). All from *Burney Newspaper Collection* (as earlier).

24. On lithotomy, see Ellis, *The Cambridge Illustrated History of Surgery*, pp. 180–94.

25. Wilson, *Surgery, Skin and Syphilis*, pp. 45–50.

26. Wiseman, *Several Chirurgical Treatises*, p. 113.

27. Ibid., p. 117.

28. Ibid. See also Dionis, *A Course of Chirurgical Operations*, pp. 249–50.

29. Hippocrates and Celsus (Philippus Aureolus Theophrastus Bombastus von Hohenheim), *The Aphorisms of Hippocrates, and the Sentences of Celsus* (trans. with additions by C.J. Sprengell) (London: 1708), p. 170.

30. Théophile Bonet, *A Guide to the Practical Physician* (London: 1684), p. 61.
31. Dionis, *A Course of Chirurgical Operations*, pp. 256–7; my italics.
32. Johannes Scultetus, *The Chyrurgeons Store-House* (trans. 'E.B.') (London: 1674), sig. A4r.
33. Ibid.
34. For removal of the eye, see John Pechey, *The Store-House of Physical Practice: Being a General Treatise of the Causes and Signs of all Diseases Afflicting Human Bodies* (London: 1695), p. 63; for removal of a (probably venereal) tumour of the scrotum, culminating in amputation of the testicles, see Hugh Ryder, *The New Practice of Chirurgery: Being a Medical Account of Divers Eminent Observations, Cases, and Cures, Very Necessary and Useful for Surgeons in the Military and Naval Service* (second edition) (London: 1693 [1689]), pp. 55–8.
35. Wiseman, *Several Chirurgical Treatises*, p. 103.
36. Alexander Read and unknown author, *Chirurgorum Comes: or The Whole Practice of Chirurgery* (London: 1687), p. 26.
37. Wilson, *Surgery, Skin and Syphilis*, especially pp. 60–2; Olivia Weisser, '(Roy Porter Student Prize Essay) Boils, Pushes and Wheals: Reading Bumps on the Body in Early Modern England', *Social History of Medicine* 22:2 (2009), 326 (quoted).
38. Wiseman, *Several Chirurgical Treatises*, p. 111.
39. Ibid.
40. Ibid., pp. 111–12.
41. Ibid., p.115.
42. Ibid.
43. Ibid., pp. 115–16. The *Oxford English Dictionary* lists 'alveoli' (sing. *alveolus*) as referring at this time to 'The cavity in a jawbone in which the root of a tooth is contained; a tooth socket'. 'alveolus, n.'. OED Online, March 2013, Oxford University Press, http://www.oed.com, 30 May 2013.
44. Wiseman, *Several Chirurgical Treatises*, p. 116.
45. Ambroise Paré, *The Workes of that Famous Chirurgion Ambrose Parey* (trans. Thomas Johnson, book 29 trans. George Baker) (London: 1634), p. 281.
46. Ibid., pp. 281–2.
47. See Henri-François Le Dran, *Observations in Surgery* (trans. J[ohn] S[parrow]) (London: 1739), pp. 43–6; Bonet, *A Guide to the Practical Physician*, p. 63.
48. William Beckett, *New Discoveries Relating to the Cure of Cancers .. To Which is Added, a Solution of some Curious Problems, Concerning the Same Disease* (1711), pp. 69–70.
49. Paul Barbette, with Raymundus Minderius (Raymond Minderer), *Thesaurus Chirurgiae: The Chirurgical and Anatomical Works of Paul Barbette, M.D., Practitioner at Amsterdam* (fourth edition) (London: 1687), p. 125.
50. Dionis, *A Course of Chirurgical Operations*, p. 254.
51. Ibid., pp. 254–5.
52. Kaartinen, *Breast Cancer in the Eighteenth Century*, p. 49.
53. Joseph Binns, *Casebook*, British Library Sloane MS.153, p. 241.
54. Dionis, *A Course of Chirurgical Operations*, p. 253.
55. Beckett, *New Discoveries*, p. 46.
56. Removal of the muscle and lymphatic structures underlying the breast became popular in the later eighteenth century. See Kaartinen, *Breast Cancer in the Eighteenth Century*, pp. 52–3; Ellis, *The Cambridge Illustrated History of Surgery*, p. 170.

57. Bonet, *A Guide to the Practical Physician*, p. 62.
58. Dionis, *A Course of Chirurgical Operations*, p. 255.
59. Kaartinen, *Breast Cancer in the Eighteenth Century*, pp. 51–2.
60. Handley, *Colloquia Chirurgica*, pp. 69–70.
61. Ibid.
62. Wiseman, *Several Chirurgical Treatises*, p. 107.
63. Ibid.
64. Dionis, *A Course of Chirurgical Operations*, p. 255.
65. T.D., *The Present State of Chyrurgery, with Some Short Remarks on the Abuses Committed Under a Pretence to the Practice* (London: 1703), pp. 19–20.
66. Philip Barrough, *The Method of Physick Conteyning the Causes, Signes, and Cures of Inward Diseases in Mans Body from the Head to the Foote* (London: 1583), p. 232.
67. Jacques Guillemeau, 'A.H.' and W. Bailey, *A Worthy Treatise of the Eyes .. Togeather With a Profitable Treatise of the Scorbie; & Another of the Cancer by A. H.* (London: 1587), p. 64.
68. Wilson, *Surgery, Skin and Syphilis*, pp. 17–20.
69. Demaitre, 'Medieval Notions of Cancer', 630–2.
70. Ibid., 102–17.
71. Ward, *Diary of the Rev. John Ward*, pp. 254–7.
72. Sawday, *The Body Emblazoned*, especially pp. 45–50; French, *Dissection and Vivisection*, especially pp. 2–7; Egmond, 'Execution, Dissection, Pain and Infamy'.
73. Sawday, *The Body Emblazoned*, p. 49.
74. Egmond, 'Execution, Dissection, Pain and Infamy', p. 114; French, *Dissection and Vivisection*, p. 2. Closely related to this fear was the alignment of surgeons with torturers: see Lisa Silverman, *Tortured Subjects: Pain, Truth and the Body in Early Modern France* (Chicago: The University of Chicago Press, 2001), pp. 133–47.
75. Richard Sugg, *Murder after Death: Literature and Anatomy in Early Modern England* (Ithaca, NY: Cornell University Press, 2007), p. 164.
76. Michael Drayton, 'Sonnet 50' from *Ideas* (1605). Reprinted in J. William Hebel (ed.), *The Works of Michael Drayton*, 3 vols. (Oxford: Basil Blackwell, 1932), p. 335.
77. Sugg, *Murder after Death*, p. 191.
78. Andrew Wear, 'Medical Ethics in Early Modern England', in Andrew Wear, Johanna Geyer-Kordesch and Roger French (eds), *Doctors and Ethics: The Earlier Historical Setting of Professional Ethics* (Amsterdam: Rodopi, 1993), pp. 98–130; Pouchelle, *The Body and Surgery*, p. 76. See also: Seth Stein LeJacq, 'The Bounds of Domestic Healing: Medical Recipes, Storytelling and Surgery in Early Modern England', *Social History of Medicine* 26:3 (2013), 451–68.
79. John Woodall, *The Surgions Mate* (London: 1617), p. 6.
80. Wiseman, *Several Chirurgical Operations*, pp. 108–9.
81. Ibid.
82. Scultetus, *The Chyrurgeons Store-House*, p. 171; my italics.
83. Ibid.
84. Ibid.
85. Payne, *With Words and Knives*, pp. 108–13.
86. Robert James, 'amputation', in *A Medicinal Dictionary*, 3 vols. (London, 1743–1746), vol. 1, sig. 5Gv.

87. Ibid.
88. Payne, *With Words and Knives*, p. 87.
89. Dionis, *A Course of Chirurgical Operations*, pp. 254–5.
90. T.D., *The Present State of Chyrurgery*, pp. 19–20.
91. Ibid.
92. Ibid.
93. Ibid.
94. Beckett, *New Discoveries*, pp. 8–9.
95. John Donne, 'Elegy 14: The Comparison', from *John Donne: The Major Works*, ed. John Carey (Oxford; New York: Oxford University Press, 1990), pp. 62–3.
96. Roy Porter, *Bodies Politic: Disease, Death and Doctors in Britain, 1650–1900* (London: Reaktion Books, 2001), p. 222.
97. Ibid.
98. Kaartinen, *Breast Cancer in the Eighteenth Century*, especially pp. 67–78.
99. Laura Gowing, *Common Bodies: Women, Touch and Power in Seventeenth-Century England* (New Haven; London: Yale University Press, 2003), p. 53.
100. Ibid., p. 16.
101. John Pechey, Theodore Mayern (Sir Théodore Turquet de Mayerne), Dr. Chamberlain (probably Thomas Chamberlayne) and Nicholas Culpeper, *The Compleat Midwife's Practice* (London: 1698), p. 186. See the bibliography for details of the multiple editions and authorship of this text.
102. Ibid.
103. George Ballard, *Memoirs of Several Ladies of Great Britain, Who have been Celebrated for their Writings or Skill in the Learned Languages, Arts, and Sciences* (Oxford: 1752), pp. 445–60. When she consented to mastectomy, in 1731, Astell reportedly insisted on there being as few people as possible present at the operation.
104. *An Account of the Causes of Some Particular Rebellious Distempers viz. the Scurvey, Cancers in Women's Breasts, &c. Vapours, and Melancholy, &c.* (London: 1670), p. 24.
105. Daniel Turner, *De Morbis Cutaneis: Diseases Incident to the Skin* (London: 1714), p. 127.
106. Ibid.
107. The *Oxford English Dictionary* identifies the first use of 'serve' to describe acting as 'the servant or lover of (a lady)' in *c*.1374, and its use 'Of a male animal: To cover (the female)' in 1577. 'serve, v.1'. OED Online, http://www.oed.com, 8 February 2013.
108. Dionis, *A Course of Chirurgical Operations*, pp. 256–7.
109. Ibid.
110. Donegan and Spratt, *Cancer of the Breast*, p. 3.
111. Ellen Leopold, *A Darker Ribbon: Breast Cancer, Women, and Their Doctors in the Twentieth Century* (Boston: Beacon Press, 1999), especially pp. 5 and 48–70; Barron H. Lerner, *The Breast Cancer Wars: Hope, Fear, and the Pursuit of a Cure in Twentieth-Century America* (New York: Oxford University Press, 2001).
112. Bridget L. Goodbody, '"The Present Opprobrium of Surgery": "The Agnew Clinic" and Nineteenth-Century Representations of Cancerous Female Breasts', *American Art* 8:1 (1994), 48.

113. Ibid.
114. Leopold, *A Darker Ribbon*, p. 5.
115. *An Account of the Causes of Some Particular Rebellious Distempers*, pp. 24–5.
116. F. David Hoeniger, *Medicine and Shakespeare in the English Renaissance* (Newark: University of Delaware Press, 1992), p. 104.
117. *An Account of the Causes of Some Particular Rebellious Distempers*, pp. 24–5.
118. Paster, *The Body Embarrassed*, especially Chapter 2, 'Laudable Blood: Bleeding, Difference, and Humoral Embarrassment', pp. 64–112.
119. Sawday, *The Body Emblazoned*, p. 3.
120. See: Roger Lund, 'Infectious Wit: Metaphor, Atheism and the Plague in Eighteenth-Century London', *Literature and Medicine* 22:1 (2003), 45–64; Lucinda Cole, 'Of Mice and Moisture: Rats, Witches, Miasma, and Early Modern Theories of Contagion', *Journal for Early Modern Cultural Studies* 10:2 (2010), 65–84; Donald Beecher, 'An Afterword on Contagion', in Claire L. Carlin (ed.), *Imagining Contagion in Early Modern Europe* (Basingstoke: Palgrave Macmillan, 2005), pp. 243–60; Jon Arrizabalaga, John Henderson and Roger French, *The Great Pox: The French Disease in Renaissance Europe* (New Haven; London: Yale University Press, 1997), pp. 245–7.
121. Eric Langley, 'Plagued by Kindness: Contagious Sympathy in Shakespearean Drama', *Medical Humanities* 37 (2011), 104.
122. Barbara M. Benedict, *Curiosity: A Cultural History of Early Modern Inquiry* (Chicago: University of Chicago, 2001), p. 2.
123. Paster, *The Body Embarrassed*, p. 237.

Conclusion: 'Death Is Only Their Desire'

Richard Wiseman, *Several Chirurgical Treatises* (second edition) (London: 1686), p. 117.

1. Reverend John Ward, *Diary of the Rev. John Ward, A.M., Extending from 1648 to 1679*, ed. Charles Severn (London: Henry Coldurn, 1839), pp. 245–7. From *Internet Archive* (online resource), http://www.archive.org, 2 March 2012.
2. Claude Deshaies Gendron, *Enquiries into the Nature, Knowledge, and Cure of Cancers* (London: 1701), pp. 119–20; Lazarius Riverius, *The Practice of Physick* (trans. with additions by Nicholas Culpeper, Abdiah Cole and William Rowland) (London: 1655), pp. 492–3; Daniel Sennert, Nicholas Culpeper and Abdiah Cole, *Practical Physick: The Fourth Book, in Three Parts* (London: 1664), p. 215.
3. Riverius, *The Practice of Physick*, pp. 492–3. See also Lazarius Riverius, *Four Books of That Learned and Renowned Doctor, Lazarus Riverius*. Appended to Felix Platter, Abdiah Cole and Nicholas Culpeper, *A Golden Practice of Physick* (London: 1662), p. 83; image 624; John Pechey, *A General Treatise of the Diseases of Maids, Bigbellied Women, Child-Bed-Women, and Widows Together with the Best Methods of Preventing or Curing the Same* (London: 1696), p. 224; Théophile Bonet, *A Guide to the Practical Physician* (London: 1684), p. 62.
4. See Michael Stolberg, *Experiencing Illness and the Sick Body in Early Modern Europe* (Basingstoke: Palgrave Macmillan, 2011).

5. *An Account of the Causes of Some Particular Rebellious Distempers viz. the Scurvey, Cancers in Women's Breasts, &c. Vapours, and Melancholy, &c.* (London: 1670), p. 25.
6. Ambroise Paré, *The Case Reports and Autopsy Records of Ambroise Paré*, ed. Wallace B. Hamby (trans. from J.P. Malgaigne (ed.)', 'Peuvres Completes d'Ambroise Paré' [Paris, 1840]) (USA: Charles C. Thomas, 1960), pp. 15–16.
7. *An Account of the Damnable Prizes in Old Nicks Lottery, for Men of Honour Only; Where Every Man that Ventures, is Sure to Get the Lord Knows What For Ever* (London: 1712), p. 3.
8. Gideon Harvey, *Little Venus Unmask'd* (seventh edition) (London: 1702 (1670)), pp. 70–1.
9. On this subject, see Michael Stolberg, 'Active Euthanasia in Pre-Modern Society, 1500–1800: Learned Debates and Popular Practices', *Social History of Medicine* 20:2 (2007), 205–21.
10. See Barron H. Lerner, *The Breast Cancer Wars: Hope, Fear, and the Pursuit of a Cure in Twentieth-Century America* (New York: Oxford University Press, 2001), especially pp. 269–74.
11. Hephzibah Roskelly, 'I Meditate on Descartes', *Social Semiotics* 22:1 (February 2012), 35. See also Nadine Ehlers and Shiloh Krupar, 'Introduction: The Body in Breast Cancer', *Social Semiotics* 22:1 (February 2012), 1–11; James T. Patterson, *The Dread Disease: Cancer and Modern American Culture* (Cambridge, MA: Harvard University Press, 1987), especially pp. 30–2.

Bibliography

Primary Bibliography

Printed Sources

A Physical Dictionary. Or, an Interpretation of such Crabbed Words and Terms of Art, as are Deriv'd from the Greek or Latin, and Used in Physick, Anatomy, Chirurgery, and Chymistry (London: 1657).

A.M., *A Rich Closet of Physical Secrets, Collected by the Elaborate Paines of Four Severall Students in Physick, and Digested Together; viz., The Child-bearers Cabinet. A Preservative Against the Plague and Small Pox. Physicall Experiments Presented to our Late Queen Elizabeths Own Hands. With Certain Approved Medicine, Taken out of a Manuscript, Found at the Dissolution of one of our English Abbies, and Supplied with Some of his Own Experiments, by a Late English Doctor* (London: 1652).

A.T., *A Rich Store-House or Treasury for the Diseased. Wherein, are Many Approved Medicines for Divers and Sundry Diseases, which have been Long Hidden, and Not Come to Light Before this Time* (London: 1596).

Adams, Thomas, *The Blacke Devil or the Apostate. Together with the Wolfe Worrying the Lambes and The Spirituall Navigator, Bound for the Holy Land* (London: 1615).

Ainsworth, Robert (edited and abridged by 'Mr. Thomas'), *An Abridgement of Ainsworth's Dictionary of the Latin Tongue, from the Folio Edition... Volume 1* (London: 1758).

Amyas, Richard, *An Antidote Against Melancholy* (London: 1659).

An Account of the Causes of Some Particular Rebellious Distempers viz. the Scurvey, Cancers in Women's Breasts, &c. Vapours, and Melancholy, &c. Weaknesses in Women, &c. Gout, Fistula in Ano, Dropsy, Agues, &c. (London: 1670).

An Account of the Damnable Prizes in Old Nicks Lottery, for Men of Honour Only; Where Every Man that Ventures, is Sure to get the Lord Knows What For Ever. In a Gradation of Familiar Thoughts, Arising, upon the Not Passing of the Duelling Bill, Brought in Last Session of Parliament (London: 1712).

Aristotle, and William Salmon, *Aristotle's Compleat and Experienc'd Midwife: In Two Parts* (London: 1711). Though this text is attributed to Aristotle, it is largely a product of the 'translator', William Salmon.

Bailey, Nathan, *An Universal Etymological English Dictionary* (London: 1721).

Ballard, George, *Memoirs of Several Ladies of Great Britain, Who Have Been Celebrated for their Writings or Skill in the Learned Languages, Arts, and Sciences* (Oxford: 1752).

Bancroft, Thomas, *Of Divine Precepts*, in *Two bookes of epigrammes, and epitaphs* (1639).

Banister, John, *An Antidotarie Chyrurgicall Containing Great Varietie and Choice of All Sorts of Medicines that Commonly Fal Into the Chyrurgions Use: Partlie Taken out of Authors, Olde and New, Printed or Written: Partlie Obtained by Free Gifte of Sundrie Worthie Men of this Profession Within this Land* (1589).

Barbette, Paul, with Raymundus Minderius (Raymond Minderer), *Thesaurus Chirurgiae: The Chirurgical and Anatomical Works of Paul Barbette, M.D., Practitioner at Amsterdam* (fourth edition) (London: 1687).

Barrough, Philip, *The Method of Physick Conteyning the Causes, Signes, and Cures of Inward Diseases in Mans Body from the Head to the Foote* (London: 1583).

Bartholin, Thomas ('published' with possible additions by Nicholas Culpeper and Abdiah Cole), *Bartholinus Anatomy Made from the Precepts of his Father, and from the Observations of All Modern Anatomists, Together with his Own* (London: 1668).

Bayfield, Robert, *Enchiridion Medicum: Containing the Causes, Signs, and Cures of All those Diseases, that Do Chiefly Affect the Body of Man: Divided Into Three Books* (London: 1655).

—— *Tractatus de Tumoribus Praeter Naturam, or, A Treatise of Preternatural Tumors* (London: 1662).

Beckett, William, *New Discoveries Relating to the Cure of Cancers, Wherein a Method of Dissolving the Cancerous Substance is Recommended, With Various Instances of the Author's Success in Such Practice, on Persons Reputed Incurable, in a Letter to a Friend. To Which is Added, a Solution of Some Curious Problems, Concerning the Same Disease* (1711).

—— *New Discoveries Relating to the Cure of Cancers. Wherein the Painful Methods of Cutting them Off, and Consuming them by Causticks are Rejected, and that of Dissolving the Cancerous Substance is Recommended, With Various Instances of the Author's Success in Such Practice, of Persons Reputed Incurable; In a Letter to a Friend to Which is Added, a Solution of Some Curious Problems, Concerning the Same Disease* (second edition) (1712).

Billingsley, Nicholas, 'On Conscience', in *A Treasury of Divine Raptures* (1667). From *English Poetry Database* (online resource), http://0-collections.chadwyck.co.uk, 3 March 2012.

Bold, Henry, 'Song XII' from *Latine Songs* (1685), pp. 44–5. From *English Poetry Database* (online resource), www.0-collections.chadwyck.co.uk, 19 February 2011.

Bonet, Théophile, *A Guide to the Practical Physician: Shewing, from the Most Approved Authors, Both Ancient and Modern, the Truest and Safest Way of Curing All Diseases, Internal and External, Whether by Medicine, Surgery, or Diet* (London: 1684).

Border, D., *Polypharmakos Kai Chymistes, or, The English Unparallel'd Physitian and Chyrurgian Shewing the True Use of All Manner of Plants and Minerals in Which is Explained the Whole Art and Secresy of Physick and Chirurgery* (London: 1651).

Borgognoni, Theodoric, *The Surgery of Theodoric* (*c.*1267) (trans. Eldridge Campbell) (New York: Appleton-Century-Crofts, Inc., 1955).

Boyle, Robert, 'Essay II. Offering some Particulars Relating to the Pathologicall Part of Physick', 'Essay III. Containing some Particulars Relating to the Semiotical Part of Physick' and 'Essay V. Proposing some Particulars Wherein Natural Philosophy May Be Useful to the Therapeutical Part of Physick', in *Some Considerations Touching the Usefulnesse of Experimental Naturall Philosophy Propos'd in Familiar Discourses to a Friend, by Way of Invitation to the Study of it* (Oxford: 1663).

Browne, John, *Adenochoiradelogia, or, An Anatomick-Chirurgical Treatise of Glandules & Strumaes or, Kings-Evil-Swellings* (London: 1684).

—— *The Surgeons Assistant ... Also a Compleat Treatise of Cancers and Gangreens. With an Enquiry Whether they have Any Alliance with Contagious Diseases* (London: 1703). Imperfect pagination.

Brownrigg, William, 'The Medical Casebook of William Brownrigg, MD, FRS (1712–1800) of the Town of Whitehaven in Cumberland', reproduced in *Medical History Supplement, XIII* (London: 1993).

Brugis, Thomas, *The Marrow of Physicke* (London: 1640).

Bunworth, Richard, *The Doctresse: A Plain and Easie Method, of Curing those Diseases Which are Peculiar to Women* (London: 1656).

Burton, Robert, *The Anatomy of Melancholy What it Is* (Oxford: 1621).

Butler, Samuel, 'Religion', from *Satires and Miscellaneous Poetry and Prose* (1928: the date of writing is unknown, although Butler was most active from 1650 to 1680). From *English Poetry Database* (online resource), www.0-collections.chadwyck.co.uk, 7 February 2011.

Clark, R., *Vermiculars Destroyed, with an Historical Account of Worms, Collected from the Best Authors as Well Ancient as Modern, Proved by that Admirable Invention of the Microscope* (London: 1690). An advertisement for this text shows that it was first printed in 1661, though no extant copy remains. It was reprinted at least four times until 1691.

Clowes, William, *A Short and Profitable Treatise Touching the Cure of the Disease Called Morbus Gallicus by Unctions* (1579).

The Compleat Doctoress: Or, A Choice Treatise of All Diseases Insident to Women (London: 1656).

Conway, Anne, Henry More, et al., *The Conway Letters: The Correspondence of Anne, Viscountess Conway, Henry More, and their Friends, 1642–1684*, ed. Marjorie Hope Nicolson. Revised by Sarah Hutton [New York: Oxford University Press, 1992 (1930)].

Cotton, Charles, 'Contentment: Pindarick Ode', in John Beresford (ed.), *Poems of Charles Cotton, 1630–1687* (London: Cobden-Sanderson, 1923), pp. 224–8.

Cowper, Sarah, *Diary* (1700–03), *Defining Gender* (online resource), http://www.amdigital.co.uk/m-collections/collection/defining-gender-1450–1910/, 5 June 2009.

Culpeper, Nicholas, *A Directory for Midwives or, A Guide for Women, in their Conception, Bearing, and Suckling their Children* (London: 1651).

—— *Culpeper's Last Legacy, Left and Bequeathed to his Dearest Wife for the Publick Good, Being the Choicest and Most Profitable of those Secrets Which While he Lived were Locked Up in his Breast, and Resolv'd Never to be Published Till After his Death* (London: 1685 [first published 1655]). This text was presented as having been left by Culpeper to his wife. However, the Oxford Dictionary of National Biography notes that Alice Culpeper decried the book as an unedited version that her husband had never intended to publish. The original imprint was by Nathaniel Brook, and not Culpeper's usual printer, Peter Cole. See Patrick Curry, 'Culpeper, Nicholas (1616–1654)', *Oxford Dictionary of National Biography*, Oxford University Press, 2004, http://www.oxforddnb.com/view/article/6882, 18 June 2013.

—— *The English Physitian Enlarged with Three Hundred, Sixty, and Nine Medicines Made of English Herbs that were Not in Any Impression Until This* (London: 1653 [1652]).

Daily Gazetteer (London Edition), Issue 423 (3 November, 1736). From *Burney Newspaper Collections* (online resource), http://0-find.galegroup.com, 4 March 2013.

Daily Journal Issue 2652 (London: 8 July 1729). From *Burney Newspaper Collections* (online resource), http://0-find.galegroup.com, 4 March 2013.

de Chauliac, Guy, *Grande Chirurgerie*, ed. E. Nicaise (Paris, 1890 [1363]).

de Voragine, Jacobus (selected and trans. Christopher Stace, with an introduction and notes by Richard Hamer), *The Golden Legend: Selections* (Harmondsworth: Penguin, 1998).

Deshaies Gendron, Claude, *Enquiries Into the Nature, Knowledge, and Cure of Cancers* (London: 1701).

Dionis, Pierre, *A Course of Chirurgical Operations, Demonstrated in the Royal Garden at Paris* (London: 1710 (French edition 1707).

—— *A General Treatise of Midwifery* (London: 1719).

'Doctors Find Worm in Woman's Brain While Operating on "Tumour"', *Daily Mail* (20 November 2008), http://www.dailymail.co.uk/news/article-1087937/ Doctors-worm-womans-brain-operating-tumour.html, 3 February 2012.

Donne, John, 'Elegy 14: The Comparison', in John Carey (ed.), *John Donne: The Major Works* (Oxford; New York: Oxford University Press, 1990), pp. 62–3.

Drayton, Michael, 'Sonnet 50', from *Ideas* (1605). Reprinted in J. William Hebel (ed.), *The Works of Michael Drayton*, 3 vols. (Oxford: Basil Blackwell, 1932), p. 335.

Dubé [D'Ube], Paul, *The Poor Man's Physician and Surgeon: Shewing the True Method of Curing All Sorts of Distempers, by the Help of Such Medicines as are of the Product of Our Climate, and Consequently to be Prepared Without Much Charge and Difficulty* (London: 1704).

D'Urfey, Thomas, *Madam Fickle: Or The Witty False One. A Comedy* (London: 1677).

Edgeworth, Maria, *Belinda* (1801) (New York: Oxford University Press, 1996).

Edwards, Edward, *The Whole Art of Chirurgery* (London: 1637).

The English Midwife Enlarged Containing Directions to Midwives; Wherein is Laid Down Whatever is Most Requisite for the Safe Practising her Art (London: 1682).

Every Woman Her Own Midwife: Or a Compleat Cabinet Opened for Child-Bearing Women. Furnished with Directions to Prevent Miscarriages During the Time of Breeding, and Other Casualties which Usually Attend Women in Child-Bed: To Which is Annexed Cures for All Sorts of Diseases Incident to the Bodies of Men, Women and Children, to Which is Appended Choise and Select Medicines, Collected by a Phisitian for his Own Private Use, and Alphabetically Digested by him, and from him Communicated for Publick Use (London: 1675).

The Examiner; Or, Remarks Upon Papers and Occurrences 6:16 (16 July 1714). From *Eighteenth Century Journals* (online resource), http://0-www.18thcjournals. amdigital.co.uk, 28 September 2012.

Fernelius, John, *Select Medicinal Counsels of John Fernelius, Chief Physitian to the King*. Appended to Felix Platter, Abdiah Cole and Nicholas Culpeper, *A Golden Practice of Physick* (London: 1662). Two versions: this is Wing 1996:7b. This is a complex composite text composed of re-editions and direct copies of several books from different authors. Parts of the book are separately paginated, therefore for clarity, I have given the image number for the EEBO digitised text alongside the page number in each reference in the endnotes.

Fletcher, John, *The Faithful Shepherdess* (1608). From *English Drama* (online resource), www.0-collections.chadwyck.co.uk, 30 September 2012.

Fontanus, Nicholas (Nicholaas Fonteyn), *The Womans Doctour* (London: 1652).

Gabelkover, Oswald, *The Boock of Physicke* (trans. 'A.M.') (Dorte [probably modern Dortmund]: 1599).

Galen of Pergamon, *Certaine Workes of Galens, Called Methodus Medendi, with a Briefe Declaration of the Worthie Art of Medicine, the Office of a Chirurgion, and an Epitome of the Third Booke of Galen, of Naturall Faculties* (trans. Thomas Gale) (London: 1566).

—— (ed. and trans., with an introduction by Ian Johnston), *Galen: On Diseases and Symptoms* (Cambridge: Cambridge University Press, 2006).

Gibson, Thomas, *The Anatomy of Humane Bodies Epitomized Wherein All Parts of Man's Body, with their Actions and Uses, are Succinctly Described, According to the Newest Doctrine of the Most Accurate and Learned Modern Anatomists* (London: 1682).

Grey, Elizabeth, Countess of Kent, *A Choice Manual of Rare and Select Secrets* (London: 1653). Grey is widely credited as the author of *A Choice Manual*, but the *Oxford Dictionary of National Biography* argues that this text was compiled by William Jarvis (also the publisher) for Grey's use, with associates of Grey contributing several items, and Grey herself only one. The text was republished in at least 16 editions between 1653 and 1708. See John Considine, 'Grey, Elizabeth, Countess of Kent (1582–1651)', *Oxford Dictionary of National Biography*, Oxford University Press, 2004; online edition, October 2006, http://www.oxforddnb.com/view/article/11530, 17 June 2013.

Guillemeau, Jacques, *Childbirth, or, the Happy Delivery of Women, Wherein is Set Downe the Government of Women. In the Time of their Breeding Childe: of their Travaile, Both Naturall, and Contrary to Nature: and of their Lying In* (London: 1612 [first edition in French c.1609]).

—— *The Frenche Chirurgery, or All the Manualle Operations of Chirurgerye, With Divers, & Sundrye Figures, and Amongst the Rest, Certayne Nuefownde Instrumentes, Verye Necessarye to All the Operationes of Chirurgerye* (trans. 'A.M.') (Dort[mund]: 1598).

—— With 'A.H.', and W. Bailey, *A Worthy Treatise of the Eyes … Togeather With a Profitable Treatise of the Scorbie; & Another of the Cancer by A. H.* (London: 1587).

Hall, John, *Select Observations on English Bodies, or, Cures Both Empericall and Historicall Performed Upon Very Eminent Persons in Desperate Diseases* (London: 1657).

Handley, James, *Colloquia Chirurgica: Or, the Whole Art of Surgery Epitomiz'd and Made Easie, According to Modern Practice* (London: 1705).

Hartman, G., *The Family Physitian, or, A Collection of Choice, Approv'd and Experienced Remedies, for the Cure of Almost All Diseases Incident to Human Bodies, Whether Internal or External; Useful in Families, and Very Serviceable to Country People* (London: 1696).

Harvey, Gideon, *Little Venus Unmask'd* (seventh edition) [London: 1702 (1670)].

Hippocrates and Celsus (Philippus Aureolus Theophrastus Bombastus von Hohenheim), *The Aphorisms of Hippocrates, and the Sentences of Celsus; With Explanations and References to the Most Considerable Writers in Physick and Philosophy, Both Ancient and Modern*, ed. C.J. Sprengell (London: 1708).

Hobbes, Thomas, *Leviathan, or, The Matter, Form, and Power of a Common-Wealth Ecclesiastical and Civil* (London: 1651).

Holden, Mary, *The Woman's Almanack or Ephemerides for the Year of Our Lord, 1689* (1689).

Isham, Elizabeth, *Booke of Rememberance* (c.1638), from *Constructing Elizabeth Isham*, University of Warwick, http://web.warwick.ac.uk/english/perdita/Isham, 7 April 2010.

Jones, [George?], *Jones of Hatton-Garden, his Book of Cures* (London: 1673).

Jonstonus, Johannes, *The Idea of Practicall Physick* (trans. Nicholas Culpeper) (London: 1657).

Keach, Benjamin, 'Hymn 146: No Light, But Darkness There Doth Dwell', from *Spiritual Melody* (1691).

Lanfranco, Giovanni (Lanfranco of Milan), *A Most Excellent and Learned Woorke of Chirurgerie, Called Chirurgia Parua Lanfranci Lanfranke of Mylayne his Briefe* (trans. John Halle) (London: 1565).

Le Clerc, Charles Gabriel, (with appended texts by M. Duncan and M. Arnaud), *The Compleat Surgeon: Or, the Whole Art of Surgery Explain'd in a Most Familiar Method* (London: 1701).

—— *A Description of Bandages and Dressings, According to the Most Commodious Ways Now Used in France* (London: 1701).

Le Clerc, Daniel, *A Natural and Medicinal History of Worms, Bred in the Bodies of Men and Other Animals* (London: 1721). From *Open Library* (online resource), http://openlibrary.org/books, 26 April 2013.

Le Dran, Henri-François, *Observations in Surgery: Containing One Hundred and Fifteen Different Cases with Particular Remarks on Each, for the Improvement of Young Students* (trans. J[ohn] S[parrow]) (London: 1739).

Levens, Peter, *The Path-Way to Health: Wherein are to be Found Most Excellent and Approved Medicines of Great Virtue, as Also Notable Potions and Drinks, With the Art of Distilling Divers Precious Water, for Making of Oyls, and Other Comfortable Receits for the Health of the Body, Never Before Printed* (London: 1654 [first published 1587]).

'Live Worm "Burrowed Through Woman's Brain"', *Nine MSN* (21 November 2008), http://news.ninemsn.com.au/article.aspx?id=669745, 3 February 2013.

London Evening Post. From *Burney Newspaper Collection* (online resource), http://0-find.galegroup.com, 4 March 2013

—— Issue 27 (London: 8–10 February 1728).

—— Issue 135 (London: 17–19 October 1728).

—— Issue 187 (15–18 February 1729).

Lowe, Peter, *The Whole Course of Chirurgerie, Wherein is Briefly Set Downe the Causes, Signes, Prognostications & Curations of All Sorts of Tumors, Wounds, Ulcers, Fractures, Dislocations & All Other Diseases, Usually Practiced by Chirurgions, According to the Opinion of All Our Auncient Doctours In Chirurgerie* (London: 1597).

Lower, Richard, *Dr. Lower's, and Several Other Eminent Physicians Receipts: Containing the Best and Safest Method For Curing Most Diseases in Humane Bodies* (second edition) (London: 1701).

Malynes, Gerrard, *A Treatise of the Canker of Englands Common Welth* (London: 1601).

Marten, John, *Gonosologium Novum: Or, a New System of All the Secret Infirm and Diseases, Natural, Accidental, and Venereal in Men and Women, that Defile and*

Ruin the Healths of Themselves and Their Posterity, Obstruct Conjugal Delectancy and Pregnancy, With their Various Methods of Cure (London: 1709).

Marvell, Andrew, 'To His Coy Mistress', in David Norbrook and H.R Woudhuysen (eds), *The Penguin Book of Renaissance Verse, 1509–1659* (London; New York: Penguin, 1992), pp. 372–3.

Massaria, Alessandro, *De Morbis Foemineis, the Womans Counsellour: Or, The Feminine Physitian* (trans. 'R.T.') (London: 1657).

Maynwaringe, Everard, *The Frequent, But Unsuspected Progress of Pains, Inflammations, Tumors, Apostems, Ulcers, Cancers, Gangrenes and Mortifications, Internal* (London: 1679).

Milton, 'Arcades', in *Milton: Complete Shorter Poems* (second edition), ed. John Carey (New York: Addison Wesley Longman, Inc., 1997), pp. 161–6.

Moffet, Thomas, *The Theater of Insects, or, Lesser Living Creatures, appended to Edward Topsell, The History of Four-Footed Beasts and Serpents* (London: 1658). (*The Theater of Insects* is a translation from the Latin original *Insectorum, sive, Minimorum Animalium Theatrum* [1634]).

Motteville, Madame de, *Memoirs of Madame de Motteville on Anne of Austria and Her Court*, 3 vols., ed. C.A. Saint-Beuve (Whitefish, MT: Kessinger Publishing, 2005), vol. 3.

Moyle, John, *The Experienced Chirurgion* (London: 1703).

The Original Weekly Journal: With Fresh Advices, Foreign and Domestick (11–18 May 1717). From *Eighteenth Century Journals* (online resource), http://0-www.18thcjournals.amdigital.co.uk, 28 September 2012.

Paré, Ambroise, *The Case Reports and Autopsy Records of Ambroise Paré*, ed. Wallace B. Hamby [trans. from J.P Malgaigne (ed.), 'Peuvres Completes d'Ambroise Paré' (Paris, 1840)] (USA: Charles C. Thomas, 1960).

—— *The Workes of that Famous Chirurgion Ambrose Parey Translated Out of Latine and Compared with the French* (trans. Thomas Johnson, book 29 trans. by George Baker) (London: 1634). *The Workes* is a translated collection of several of Paré's texts dating from the mid-late sixteenth century, first published in France in 1575.

Partridge, John, *The Widowes Treasure Plentifully Furnished with Sundry Precious and Approoved Secretes in Phisicke and Chirurgery for the Health and Pleasure of Mankinde: Hereunto are Adjoyned, Sundry Pretie Practises and Conclusions of Cookerie: With Many Profitable and Holesome Medicines for Sundrie Diseases in Cattell* (London: 1588).

Pechey, John, *A General Treatise of the Diseases of Maids, Bigbellied Women, Child-Bed-Women, and Widows Together With the Best Methods of Preventing or Curing the Same* (London: 1696).

——*The Store-House of Physical Practice: Being a General Treatise of the Causes and Signs of All Diseases Afflicting Human Bodies* (London: 1695).

Pechey, John, Theodore Mayern (Sir Théodore Turquet de Mayerne), Dr. Chamberlain (probably Thomas Chamberlayne) and Nicholas Culpeper, *The Compleat Midwife's Practice Enlarged in the Most Weighty and High Concernments of the Birth of Man Containing a Perfect Directory or Rules for Midwives and Nurses* (London: 1698). (This text is the fourth edition of a number appearing under this authorship from 1659. However, Helen King identifies the text as closely related to the 1656 *The Compleat Midwife's Practice* by four midwives, 'T.C.', 'I.D.', 'M.S.' and 'T.B.'. The content is altered in the later books, but the preface remains the same, including a complaint about Culpeper's 'imperfect' work. See Helen King, *Midwifery, Obstetrics and the Rise of Gynaecology: The Uses of a Sixteenth-Century Compendium* (Aldershot; Burlington, VT: Ashgate, 2007), p. 21.

Pepys, Samuel, *Diary*. From *The Diary of Samuel Pepys* (online resource), http://www.pepysdiary.com, 5 February 2013.

Post Boy, Issue 4008 (London: 7–9 April 1715). From *Burney Newspaper Collection* (online resource), http://0-find.galegroup.com, 4 March 2013.

Purmann, Matthias Gottfried (with appended text by Conrade Joachim Sprengell), *Chirurgia Curiosa: Or, The Newest and Most Curious Observations and Operations in the Whole Art of Chirurgery ... To Which is Added Natura Morborum Medicatrix: Nature Cures Diseases* (London: 1706).

Ramesey, William, *Helminthologia, or, Some Physical Considerations of the Matter, Origination, and Several Species of Wormes Macerating and Direfully Cruciating Every Part of the Bodies of Mankind* (London: 1668).

Read, Alexander, *The Chirurgicall Lectures of Tumors and Ulcers* (London: 1635).

—— *Most Excellent and Approved Medicine Lately Compiled and Extracted Out of the Originals of the Most Famous and Best Experienced Physicians Both in England and Other Countries, by A. R. Doctor in Physick Deceased* (London: 1651).

—— With unknown author, *Chirurgorum Comes: Or The Whole Practice of Chirurgery. Begun by the Learned Dr. Read; Continued and Completed by a Member of the College of Physicians in London* (London: 1687).

—— *The Workes of that Famous Physician Dr. Alexander Read* (second edition) (London: 1650).

Riolan, Jean, *A Sure Guide, or, The Best and Nearest Way to Physick and Chyrurgery that is to Say, the Arts of Healing by Medicine and Manual Operation, Being an Anatomical Description of the Whol Body of Man and its Parts: With their Respective Diseases Demonstrated from the Fabrick and Use of the Said Parts* (trans. Nicholas Culpeper and 'W.R.') (London: 1657).

Riverius, Lazarius, *Four Books of that Learned and Renowned Doctor, Lazarus Riverius*. Appended to Felix Platter, Abdiah Cole and Nicholas Culpeper, *A Golden Practice of Physick* (London: 1662). *Two versions: this is Wing 1996:7b*. This is a complex composite text composed of re-editions and direct copies of several books from different authors. Puzzlingly, this version of *Four Books* contains observations not printed in later versions of that text. Parts of the book are separately paginated, therefore for clarity, I have given the image number for the EEBO digitised text alongside the page number in each reference in the footnotes.

—— *The Practice of Physick in Seventeen Several Books Wherein is Plainly Set Forth the Nature, Cause, Differences, and Several Sorts of Signs: Together With the Cure of all Diseases in the Body of Man* (trans. with additions by Nicholas Culpeper, Abdiah Cole and William Rowland) (London: 1655). Material from this text is substantially repeated in *Four Books*.

Ryder, Hugh, *The New Practice of Chirurgery: Being a Medical Account of Divers Eminent Observations, Cases, and Cures, Very Necessary and Useful for Surgeons in the Military and Naval Service* (second edition) (London: 1693 [1689]).

Sadler, John, *The Sicke Woman's Private Looking-Glasse. Wherein Methodically Are Handled All Uterine Affects or Diseases Arising from the Womb. Enabling Women to Informe the Physitian About the Cause of their Griefe* (London: 1636).

Salmon, William, *Paraieremata, or Select Physical and Chirurgical Observations: Containing Divers Remarkable Histories of Cures, Done by Several Famous Physicians* (London: 1687).

Scultetus, Johannes, *The Chyrurgeons Store-House* (trans. 'E.B.') (London: 1674).

Sennert, Daniel, Nicholas Culpeper and Abdiah Cole, *Practical Physick: The Fourth Book, in Three Parts* (London: 1664). (Part of a series of 30 titles by various authors called 'The Rationall Physitian's Library' or 'The Physitian's Library'.)

Sermon, William, *The Ladies Companion, or, The English Midwife Wherein is Demonstrated the Manner and Order How Women Ought to Govern Themselves During the Whole Time of their Breeding Children and of their Difficult Labour, Hard Travail and Lying-in, etc.* (London: 1671).

Shakespeare, William, *The Complete Works* (second edition), ed. Stanley Wells, Gary Taylor, John Jowett, and William Montgomery (Oxford: Oxford University Press, 2005).

Sharp, Jane, *The Midwives Book, or, The Whole Art of Midwifery Discovered, Directing Childbearing Women How to Behave Themselves in their Conception, Bearing and Nursing of Children in Six Books* (London: 1671).

Smith, Henry, 'The Benefit of Contentation' in *The Sermons of Maister Henrie Smith Gathered Into One Volume. Printed According to his Corrected Copies in his Lifetime* (London: 1593), p. 209.

Smith, John, *A Compleat Practice of Physic* (London: 1656).

The St. James's Evening Post 288 (12 March 1717). From *Eighteenth Century Journals* (online resource), http://0-www.18thcjournals.amdigital.co.uk, 28 September 2012.

T.D., *The Present State of Chyrurgery, With Some Short Remarks on the Abuses Committed Under a Pretence to the Practice. And Reasons offer'd for Regulating the Same. In a Letter to Charles Bernard, Esq; Serjeant-Surgeon; and Chyrurgeon in Ordinary to Her Present Majesty* (London: 1703).

Talbot, Alethea, Countess of Arundel, *Natura Exenterata: or Nature Unbowelled by the Most Exquisite Anatomizers of her* (London: 1655).

Tanner, John, *The Hidden Treasures of the Art of Physic* (1659).

Turner, Daniel, *De Morbis Cutaneis: Diseases Incident to the Skin* (London: 1714).

Van Helmont, Jean Baptiste, *Van Helmont's Works Containing his Most Excellent Philosophy, Physick, Chirurgery, Anatomy: Wherein the Philosophy of the Schools is Examined, their Errors Refuted, and the Whole Body of Physick Reformed and Rectified: Being a New Rise and Progresse of Philosophy and Medicine, for the Cure of Diseases, and Lengthening of Life* (trans. 'J.C.') (London: 1664).

Vicary, Thomas, *The English-Mans Treasure, With the True Anatomie of Mans Body* (ninth edition) (London: 1641).

Vigo, Giovannida, *The Most Excellent Workes of Chirurgerie Made and Set Foorthe by Maister John Vigon Head Chirurgien of Oure Tyme in Italie; Translated Into English; Wherunto is Added an Exposition of Straunge Termes and Unknowen Symples, Belongyng to the Arte* (London: 1571 [first published 1543]).

Ward, Reverend John, *Diary of the Rev. John Ward, A.M., extending from 1648 to 1679*, ed. Charles Severn (London: Henry Coldurn, 1839). From *Internet Archive* (online resource), http://www.archive.org, 2 March 2012.

Webster, John, *The White Devil*, in René Weis (ed.), *The Duchess of Malfi and Other Plays* (Oxford: Oxford University Press, 1996), pp. 1–103.

Wecker, Johann Jacob, *A Compendious Chyrurgerie: Gathered, & Translated (especially) out of Wecker, At the Request of Certaine, But Encreased and Enlightened With Certaine Annotations, Resolutions & Supplyes, Not Impertinent to this Treatise, Nor Unprofitable to the Reader* (trans. with additions by John Banister) (London: 1585).

Willis, Thomas, *Dr. Willis's Practice of Physick Being the Whole Works of that Renowned and Famous Physician Wherein Most of the Diseases Belonging to the Body of Man Are Treated Of, With Excellent Methods and Receipts for the Cure of the Same* (trans. Samuel Pordage) (London: 1684). [Trans. of *Pharmaceutice Rationalis* (1674/5)].

Winter, Salvator, *A New Dispensary of Fourty Physicall Receipts, Most Necessary and Profitable for All House-Keepers in their Families* (London: 1649).

Wirsung, Christof, *Praxis Medicinae Universalis* (trans. Jacob Mosan) (London: 1598). (Translation of *Ein New Artzney Buch, c.*1592.).

Wiseman, Richard, *Several Chirurgical Treatises* (second edition) (London: 1686).

Wither, George, 'Opobalsamum Anglicanum: An English Balme, Lately Pressed Out of a Shrub, and Spread Upon these Papers, for the Cure of Some Scabs, Gangreeves and Cancers Indangering the Bodie of this Common-Wealth', in *Miscellaneous Works* [1872–77; (*c.*1645)] from *English Poetry Database* (online resource), www.0-collections.chadwyck.co.uk, 19 February 2011.

Wolley, Hannah, *The Accomplish'd Ladies Delight in Preserving, Physic, Beautifying, and Cookery* [1686 (first published 1675)]. *The Oxford Dictionary of National Biography* states that the posthumous *Accomplish'd Ladies Delight* is an unauthorised text based on previous works by Wolley. See John Considine, 'Wolley, Hannah (*b.* 1622?, *d.* in or after 1674)', *Oxford Dictionary of National Biography*, Oxford University Press, 2004, http://www.oxforddnb.com/view/article/29957, 17 June 2013.

Wood, Owen, *An Alphabeticall Book of Physical Secrets* (London: 1639).

Woodall, John, *The Surgeons Mate* (London: 1617).

Manuscript Sources

Anon, *Collection of Medical and Cookery Receipts* (early 17th century), Wellcome Library MS.635.

Birch, Thomas, *Manuscripts Collected by Thomas Birch*, British Library Add. 4376.

Hughes, Sarah, *Mrs Hughes her Receipts* (1637), Wellcome Library MS.363.

Jones, Katherine (Viscountess Ranelagh), *Collection of Medical Receipts, c. 1675–c.1710*, Wellcome Library MS.1340.

Joseph Binns, *Casebook*, British Library Sloane MS.153.

King, Sir Edmund, *Sir Edmund King's Casebook*, 1676–96, British Library Sloane MS.1589. Pagination is irregular.

Lowdham/Loudham, Caleb and Jane Lowdham, *Notebook of Medical and Culinary Recipes with a Few Case Histories* (late 17th century – early 18th century), Wellcome Library MS.7073.

'Mrs Corylon', *A Booke of Divers Medecines* (1606), Wellcome Library MS.213.

Repp, Dorothea, *Collection of Cookery, Medical, Veterinary and Household Receipts* (early eighteenth century) Wellcome Library MS.7788.

Sleigh, Elizabeth and Felicia Whitfeld, *Collection of Medical Receipts* (1647–1722) Wellcome Library MS.751.

St John, Johanna, *Johanna St John her Booke* (1680), Wellcome Library MS.4338.

Secondary Bibliography

Ackernecht, Erwin H., 'Historical Notes on Cancer', *Medical History* 2:2 (1958), 114–19.

Albala, Ken, *Eating Right in the Renaissance* (Berkeley: University of California Press, 2002).

Amussen, Susan, *An Ordered Society: Gender and Class in Early Modern England* (New York: Columbia University Press, 1988).

Andersen, Jennifer, and Elizabeth Sauer (eds), *Books and Readers in Early Modern England: Material Studies* (Philadelphia: University of Pennsylvania Press, 2002).

Aronowitz, Robert A., *Making Sense of Illness: Science, Society and Disease* (Cambridge: Cambridge University Press, 1998).

Arrizabalaga, Jon, John Henderson and Roger French, *The Great Pox: The French Disease in Renaissance Europe* (New Haven; London: Yale University Press, 1997).

Bakhtin, Mikhail, *Rabelais and His World* (trans. Helene Iswolsky) (Bloomington: Indiana University Press, 1984).

Becker, Lucinda M., *Death and the Early Modern Englishwoman* (Aldershot: Ashgate, 2003).

Beecher, Donald, 'An Afterword on Contagion', in Claire L. Carlin (ed.), *Imagining Contagion in Early Modern Europe* (Basingstoke: Palgrave Macmillan, 2005), pp. 243–60.

Beier, Linda McCray, *Sufferers and Healers: The Experience of Illness in Seventeenth-Century England* (London: Routledge and Kegan Paul, 1987).

Benedict, Barbara M., *Curiosity: A Cultural History of Early Modern Inquiry* (Chicago; London: University of Chicago, 2001).

Bennet, Gillian, 'Bosom Serpents and Alimentary Amphibians: A Language for Sickness', in Marijke Gijswit-Hofstra, Hilary Marland and Hans de Waardt (eds), *Illness and Healing Alternatives in Western Europe* (London; New York: Routledge, 1997), pp. 224–42.

Berriot-Salvadore, Evelyne, 'The Discourse of Medicine and Science' (trans. Arthur Goldhammer), in Natalie Zemon Davis and Arlette Farge (eds), *A History of Women: Renaissance and Enlightenment Paradoxes* (Cambridge, MA: Belknap Press of Harvard University Press, 1993), pp. 348–88.

Bicks, Caroline, 'Stones Like Women's Paps: Revising Gender in Jane Sharp's *Midwives Book*', *Journal for Early Modern Cultural Studies* 7:2 (2007), 1–27.

Boehrer, Bruce, *Animal Characters: Nonhuman Beings in Early Modern Literature* (Philadelphia: University of Pennsylvania Press, 2010).

—— (ed.), *A Cultural History of Animals in the Renaissance*, Series: *A Cultural History of Animals*, 6 vols. (Oxford; New York: Berg, 2011), vol. 2.

Botelho, Lynn. 'Old Age and Menopause in Rural Women of Early Modern Suffolk', in Lynn Botelho and Pat Thane (eds), *Women and Ageing in British Society since 1500* (Harlow: Longman, 2000), pp. 43–65.

Braun, Armin C., *The Story of Cancer: On its Nature, Causes, and Control* (Reading, MA: Addison-Wesley Publishing Company, Inc., 1977).

Brockliss, Laurence, 'Medical Education and Centres of Excellence in Eighteenth-Century Europe: Towards an Identification', in Peter Ole Grell, Andrew Cunningham and Jon Arrizabalaga (eds), *Centres of Medical Excellence?*

Medical Travel and Education in Europe, 1500–1789 (Farnham: Ashgate, 2010), pp. 17–46.

Brown, Phil, 'Naming and Framing: The Social Construction of Diagnosis and Illness', *Journal of Health and Social Behavior* 35, Extra Issue: 'Forty Years of Medical Sociology: The State of the Art and Directions for the Future' (1995), 34–52.

Buckley, Thomas, and Alma Gottlieb (eds), *Blood Magic: The Anthropology of Menstruation* (Berkeley: University of California Press, 1988).

Burstein, Sona Rosa, 'Demonology and Medicine in the Sixteenth and Seventeenth Centuries', *Folklore* 67:1 (1956), 16–33.

Butler, Judith, *Gender Trouble: Feminism and the Subversion of Identity* (New York: Routledge, 1999 [1990]).

Cancer Research UK, 'Enemy' (advertisement), dir. Siri Bunford, 30 April 2013.

Capuano, James, *Beast: A Slightly Irreverent Tale About Cancer (and Other Assorted Anecdotes)* (Wickford, RI: New Street Communications LLC, 2012).

Churchill, Wendy D., *Female Patients in Early Modern Britain: Gender, Diagnosis, and Treatment* (Farnham: Ashgate, 2012).

—— 'The Medical Practice of the Sexed Body: Women, Men and Disease in Britain, circa 1600–1740', *Social History of Medicine* 18:1 (2005), 3–22.

Citrome, Jeremy J., *The Surgeon in Medieval English Literature* (New York: Palgrave Macmillan, 2006).

Clark, Stuart, *Vanities of the Eye: Vision in Early Modern European Culture* (Oxford: Oxford University Press, 2007).

Cobb, Matthew, *The Egg and Sperm Race: The Seventeenth-Century Scientists Who Unravelled the Secrets of Sex, Life and Growth* (London: The Free Press, 2006).

Cole, Lucinda, 'Of Mice and Moisture: Rats, Witches, Miasma, and Early Modern Theories of Contagion', *Journal for Early Modern Cultural Studies* 10:2 (2010), 65–84.

Comeau, Tammy Duerden, 'Gender Ideology and Disease Theory: Classifying Cancer in Nineteenth Century Britain', *Journal of Historical Sociology* 20:1/2 (2007), 158–74.

Cook, Harold J., *The Decline of the Old Medical Regime in Stuart London* (New York: Cornell University Press, 1986).

—— 'The New Philosophy and Medicine in Seventeenth-Century England', in David C. Lindberg and Robert S. Westman (eds), *Reappraisals of the Scientific Revolution* (Cambridge; New York: Cambridge University Press, 1990), pp. 397–436.

—— *Trials of an Ordinary Doctor: Joannes Groenevelt in Seventeenth-Century London* (Baltimore: Johns Hopkins University Press, 1994).

Covington, Sarah, *Wounds, Flesh and Metaphor in Seventeenth-Century England* (Basingstoke: Palgrave Macmillan, 2009).

Craik, Katharine A., *Reading Sensations in Early Modern England* (Basingstoke: Palgrave Macmillan, 2007).

Crawford, Patricia, 'Attitudes to Menstruation in Seventeenth-Century England', *Past and Present* 91 (1981), 46–73.

—— *Blood, Bodies and Families in Early Modern England* (Harlow: Pearson Longman, 2004).

—— 'Sexual Knowledge in England, 1500–1750', in Roy Porter and Mikulas Teich (eds), *Sexual Knowledge, Sexual Science: The History of Attitudes to Sexuality* (Cambridge: Cambridge University Press, 1994), pp. 82–106.

Cunningham, Andrew, *The Anatomical Renaissance: The Resurrection of the Projects of the Ancients* (Chicago: Scolar Press, 1997).

Dally, Ann, *Women Under the Knife: A History of Surgery* (London: Hutchinson Radius, 1991).

Datson, Lorraine, and Katharine Park, *Wonders and the Order of Nature, 1150–1750* (Cambridge, MA: Zone Books: Distributed by the MIT Press, 1998).

Dawson, Lesel. *Lovesickness and Gender in Early Modern English Literature* (Oxford; New York: Oxford University Press, 2008).

de Blécourt, Willem, and Cornelie Usborne, *Cultural Approaches to the History of Medicine: Mediating Medicine in Early Modern and Modern Europe* (Basingstoke: Palgrave Macmillan, 2004).

De Girolama Cheney, Liana, 'The Cult of St. Agatha', *Women's Art Journal* 17:1 (1996), 3–9.

De Moulin, Daniel, 'Historical Notes on Breast Cancer, With Emphasis on the Netherlands: I. Pathological and Therapeutic Concepts in the Seventeenth Century', *The Netherlands Journal of Surgery* 32:4 (1980), 129–34.

—— 'Historical Notes on Breast Cancer, With Emphasis on the Netherlands: II. Pathophysiological Concepts, Diagnosis and Therapy in the 18th Century', *The Netherlands Journal of Surgery* 33:4 (1981), 206–16.

—— *A Short History of Breast Cancer* (Leiden: Martinus Nijhoff Publishers, 1983).

De Renzi, Silva, 'Women and Medicine', in Peter Elmer (ed.), *The Healing Arts: Health, Disease and Society in Europe, 1500–1800* (Manchester: Manchester University Press, 2004), pp. 196–227.

—— 'Old and New Models of the Body', in Peter Elmer (ed.), *The Healing Arts: Health, Disease and Society in Europe, 1500–1800* (Manchester: Manchester University Press, 2004), pp. 166–95.

Debus, Allen G., *The Chemical Philosophy* (New York: Science History Publications, 1977).

DeLacy, Margaret, 'Nosology, Mortality, and Disease Theory in the Eighteenth Century', *Journal of the History of Medicine and Allied Sciences* 54:2 (1999), 261–84.

Delany, Paul, *British Autobiography in the Seventeenth Century* (London: Routledge, 1969).

Demaitre, Luke, *Leprosy in Premodern Medicine: A Malady of the Whole Body* (Baltimore, MD: Johns Hopkins University Press, 2007).

—— 'Medieval Notions of Cancer: Malignancy and Metaphor', *Bulletin of the History of Medicine* 72:4 (1998), 609–36.

Di Meo, Michelle, and Rebecca Laroche, 'Elizabeth Isham and Medicine', part of the *Constructing Elizabeth Isham* project headed by Elizabeth Clarke, www.warwick.ac.uk/fac/arts/ren/projects/isham/medicine, 27 January 2012.

Donegan, William L., and John Stricklin Spratt (eds), *Cancer of the Breast* (Philadelphia; London: Elsevier Science, 2002).

Dormandy, Thomas, *The Worst of Evils: The Fight Against Pain* (New Haven: Yale University Press, 2006).

Ebner, Dean, *Autobiography in Seventeenth-Century England: Theology and the Self* (Hungary: Mouton and Co., 1971).

The Economist, 'Fasting and Cancer: Starving the Beast' (article), 9 February 2012. Accessed via www.economist.com/blogs/babbage/2012/02/fasting-and-cancer, 7 October 2014.

Edwards, Karen, 'Milton's Reformed Animals: An Early Modern Bestiary: Introduction', *Milton Quarterly* 39:3 (2005), 121–31.

—— 'Milton's Reformed Animals: An Early Modern Bestiary A–C', *Milton Quarterly* 39:4 (2005), 183–292.

—— 'Milton's Reformed Animals: An Early Modern Bestiary D–F', *Milton Quarterly* 40:2 (2006), 99–187.

—— 'Milton's Reformed Animals: An Early Modern Bestiary G', *Milton Quarterly* 40:4 (2006), 263–91.

—— 'Milton's Reformed Animals: An Early Modern Bestiary H–K', *Milton Quarterly* 41:2 (2007), 79–147.

—— 'Milton's Reformed Animals: An Early Modern Bestiary L', 41:4 (2007), 223–56.

—— 'Milton's Reformed Animals: An Early Modern Bestiary M–O', *Milton Quarterly*, 42:2 (2008), 113–60.

—— 'Milton's Reformed Animals: An Early Modern Bestiary P–R', *Milton Quarterly*, 42:4 (2008), 253–308.

—— 'Milton's Reformed Animals: An Early Modern Bestiary S', *Milton Quarterly* 43:2 (2009), 89–141.

—— 'Milton's Reformed Animals: An Early Modern Bestiary T–Z', *Milton Quarterly* 43:4 (2009), 241–303.

Egmond, Florike, 'Execution, Dissection, Pain and Infamy – A Morphological Investigation', in Florike Egmond and Robert Zwijnenberg (eds), *Bodily Extremities: Preoccupations with the Human Body in Early Modern European Culture* (Aldershot: Ashgate, 2003), pp. 92–126.

Ellis, Harold, *The Cambridge Illustrated History of Surgery* (Cambridge; New York: Cambridge University Press, 2009).

Elmer, Peter, 'Chemical Medicine and the Challenge to Galenism: The Legacy of Paracelsus, 1560–1700', in Peter Elmer (ed.), *The Healing Arts: Health, Disease and Society in Europe, 1500–1800* (Manchester: Manchester University Press, 2004), pp. 108–35.

—— 'Medicine, Religion and the Puritan Revolution', in Roger French and Andrew Wear (eds), *The Medical Revolution of the Seventeenth Century* (Cambridge: Cambridge University Press, 1989), pp. 10–45.

Epstein, Julia, *The Iron Pen: Frances Burney and the Politics of Women's Writing* (Madison: University of Wisconsin Press, 1989).

—— 'Writing the Unspeakable: Fanny Burney's Mastectomy and the Fictive Body', *Representations* 16 (1986), 131–66.

Evenden, Doreen A., 'Gender Differences in the Licensing and Practice of Female and Male Surgeons in Early Modern England', *Medical History* 42 (1998), 194–216.

—— *The Midwives of Seventeenth-Century London* (New York; Cambridge: Cambridge University Press, 2000).

Farley, John, 'The Spontaneous Generation Controversy (1700–1860): The Origin of Parasitic Worms', *Journal of the History of Biology* 5:1 (1972), 95–125.

Fildes, Valerie, *Breasts, Bottles and Babies: A History of Infant Feeding* (Edinburgh: Edinburgh University Press, 1986).

—— 'The English Wet-Nurse and her Role in Infant Care 1538–1800', *Medical History* 32 (1988), 142–73.

Fitzgerald, Patrick J., *From Demons and Evil Spirits to Cancer Genes: The Development of Concepts Concerning the Causes of Cancer and Carcinogenesis* (Washington, DC: American Registry of Pathology, Armed Forces Institute of Pathology, 2000).

Forbes, Thomas R., 'Verbal Charms in British Folk Medicine', *Proceedings of the American Philosophical Society* 115:4 (1971), 293–316.

Foucault, Michel, *The Birth of the Clinic: An Archaeology of Medical Perception* (trans. A.M. Sheridan Smith) (New York: Vintage Books, 1975).

Foyster, Elizabeth, 'A Laughing Matter? Marital Discord and Gender Control in Early Modern England', *Rural History* 4:1 (1993), 5–21.

Frank Jr., Robert, 'The John Ward Diaries: Mirror of Seventeenth-Century Science and Medicine', *Journal of the History of Medicine* 29 (1974), 147–79.

French, Roger, *Dissection and Vivisection in the European Renaissance* (Aldershot; Brookfield, VT: Ashgate, 1999).

—— 'Surgery and Scrophula', in Christopher Lawrence (ed.), *Medical Theory, Surgical Practice: Studies in the History of Surgery* (London: Routledge, 1992), pp. 85–100.

Fudge, Erica, *Perceiving Animals: Humans and Beasts in Early Modern English Culture* (Basingstoke: Macmillan Press Ltd., 2000).

—— 'Renaissance Animal Things', in Joan B. Landes, Paula Young Lee and Paul Youngquist (eds), *Gorgeous Beasts: Animal Bodies in Historical Perspective* (Pennsylvania: Pennsylvania State University Press, 2012), pp. 41–56.

Furdell, Elizabeth Lane, *Publishing and Medicine in Early Modern England* (New York: University of Rochester Press, 2002).

—— *The Royal Doctors, 1485–1714: Medical Personnel at the Tudor and Stuart Courts* (New York: University of Rochester Press, 2001).

Fye, W.B., 'Active Euthanasia: A Historical Survey', *Bulletin of the History of Medicine* 52 (1978), 492–502.

Gelfland, Toby, 'Paris: 'Certainly the Best Place for Learning the Practical Part of Anatomy and Surgery'', in Peter Ole Grell, Andrew Cunningham and Jon Arrizabalaga (eds), *Centres of Medical Excellence? Medical Travel and Education in Europe, 1500–1789* (Farnham, UK; Burlington, VT: Ashgate, 2010), pp. 221–46.

Geller, Markham Judah, *Volume 7: Renal and Rectal Disease Texts* from the series *Die Babylonisch-Assyrische Medizin in Texten und Untersuchungen* (ed. Robert Biggs) (Germany: De Gruyter, 2005).

Gellert Lyons, Bridget, *Voices of Melancholy: Studies in Literary Treatments of Melancholy in Renaissance England* (London: Routledge & Kegan Paul, 1971).

Gilman, Ernest B., *Plague Writing in Early Modern England* (Chicago: University of Chicago Press, 2009).

Goodbody, Bridget L., '"The Present Opprobrium of Surgery": "The Agnew Clinic" and Nineteenth-Century Representations of Cancerous Female Breasts', *American Art* 8:1 (1994), 32–51.

Gowing, Laura, *Common Bodies: Women, Touch and Power in Seventeenth-Century England* (New Haven; London: Yale University Press, 2003).

—— *Domestic Dangers: Women, Words and Sex in Early Modern London* (Oxford: Clarendon Press, 1996).

Gowland, Angus, 'The Problem of Early Modern Melancholy', *Past and Present* 191 (2006), 77–120.

—— *The Worlds of Renaissance Melancholy: Robert Burton in Context* (Cambridge: Cambridge University Press, 2006).

Green, Monica H., 'Flowers, Poisons and Men: Menstruation in Medieval Western Europe', in Andrew Shail and Gillian Howie (eds), *Menstruation: A Cultural History* (London: Palgrave Macmillan, 2005), pp. 51–64.

—— 'From Diseases of Women to Secrets of Women', *Journal of Medieval and Early Modern Studies* 30 (2000), 5–39.

Greenblatt, Stephen. *Learning to Curse: Essays in Early Modern Culture* (New York: Routledge, 1990).

Grieco, Sara F. Matthews, 'The Body, Appearance, and Sexuality', in Natalie Zemon Davis and Arlette Farge (eds), *A History of Women: Renaissance and Enlightenment Paradoxes* (Cambridge, MA: Belknap Press of Harvard University Press, 1993), pp. 46–84.

Harley, David N., 'From Providence to Nature: The Moral Theology and Godly Practice of Maternal Breast-feeding in Stuart England', *Bulletin of the History of Medicine* 69:2 (1995), 198–223.

—— 'Medical Metaphors in English Moral Theology', *Journal of the History of Medicine and Allied Sciences* 48 (1993), 396–435.

—— 'Spiritual Physic, Providence and English Medicine, 1560–1640', in Peter Ole and Andrew Cunningham (eds), *Medicine and the Reformation* (Abingdon: Routledge, 1993), pp. 101–17.

Harley, Marta Powell, 'Last Things First in Chaucer's Physician's Tale: Final Judgment and the Worm of Conscience', *The Journal of English and Germanic Philology* 91:1 (1992), 1–16.

Harris, Jonathan Gil, '"The Canker of England's Commonwealth": Gerard Malynes and the Origins of Economic Pathology', *Textual Practice* 13:2 (1999), 311–28.

—— *Foreign Bodies and the Body Politic: Discourses of Social Pathology in Early Modern England* (Cambridge: Cambridge University Press, 1998).

—— '(Po)X Marks the Spot: How to "Read" "Early Modern" "Syphilis" in *The Three Ladies of London*', in Kevin P. Siena (ed.), *Sins of the Flesh: Responding to Sexual Disease in Early Modern Europe* (Toronto: Centre for Reformation and Renaissance Studies, 2005), pp. 109–32.

Harrison, Peter, 'The Virtues of Animals in Seventeenth-Century Thought', *Journal of the History of Ideas* 59:3 (1998), 463–84.

Healy, Margaret, 'Bodily Regimen and Fear of the Beast: "Plausibility" in Renaissance Domestic Tragedy', in Erica Fudge, Ruth Gilbert and Susan Wiseman (eds), *At the Borders of the Human: Beasts, Bodies and Natural Philosophy in the Early Modern Period* (Basingstoke: Palgrave Macmillan, 1999), pp. 51–73.

—— *Fictions of Disease in Early Modern England: Bodies, Plagues and Politics* (Basingstoke: Palgrave, 2001).

Hentschell, Roze, 'Luxury and Lechery: Hunting the French Pox in Early Modern England', in Kevin P. Siena (ed.), *Sins of the Flesh: Responding to Sexual Disease in Early Modern Europe* (Toronto: Centre for Reformation and Renaissance Studies, 2005), pp. 133–58.

Herzlich, Claude, *Illness and Self in Society* (trans. Douglas Graham) (Baltimore, MD: Academic Press Inc., 1973).

Hindson, Beth, 'Attitudes towards Menstruation and Menstrual Blood in Elizabethan England', *Journal of Social History* (2009), 89–115.

Hodges, Devon, *Renaissance Fictions of Anatomy* (New York: Cornell University Press, 1987).

Hoeniger, F. David, *Medicine and Shakespeare in the English Renaissance* (Newark: University of Delaware Press, 1992).

Huet, Marie-Hélène, *Monstrous Imagination* (Cambridge, MA: Harvard University Press, 1993).

—— 'Monstrous Medicine', in Laura Lunger Knoppers and Joan B. Landes (eds), *Monstrous Bodies/Political Monstrosities in Early Modern Europe* (Ithaca, NY: Cornell University Press, 2004), pp. 127–48.

Hunter, Lynette, 'Cankers in *Romeo and Juliet*: Sixteenth-Century Medicine at a Figural/Literal Cusp', in Stephanie Moss and Kaara L. Peterson (eds), *Disease, Diagnosis and Cure on the Early Modern Stage* (Aldershot: Ashgate, 2004), pp. 171–80.

—— 'Women and Domestic Medicine: Lady Experimenters, 1570–1620', in Lynette Hunter and Sarah Hutton (eds), *Women, Science and Medicine 1500–1700: Mothers and Sisters of the Royal Society* (Stroud: Sutton Publishing, 1997), pp. 89–107.

Iyengar, Sujata, *Shakespeare's Medical Language: A Dictionary* (London; New York: Continuum, 2011).

Jasen, P., 'Breast Cancer and the Language of Risk, 1750–1950', *Social History of Medicine* 15:1 (2002), 17–43.

Johnson, George. *The Cancer Chronicles: Unlocking Medicine's Deepest Mystery* (London: The Bodley Head, 2013).

Jordanova, Ludmilla, *Sexual Visions: Images of Gender in Science and Medicine between the Eighteenth and Twentieth Centuries* (Wisconsin: University of Wisconsin Press, 1989).

—— 'The Social Construction of Medical Knowledge', *Social History of Medicine* 8:3 (1995), 361–81.

Kaartinen, Marjo, *Breast Cancer in the Eighteenth Century* (London; Vermont: Pickering and Chatto, 2013).

Kalnin Diede, Martha, *Shakespeare's Knowledgeable Body* (New York: Peter Lang Publishing, Inc., 2008).

Kaprozilos, A., and N. Pavlidis, 'The Treatment of Cancer in Greek Antiquity', *European Journal of Cancer* 40 (2004), 2033–40.

Katriztky, M.A., *Women, Medicine and Theatre 1500–1750: Literary Mountebacks and Performing Quacks* (Aldershot; Burlington, VT: Ashgate, 2007).

Keil, Harry, 'The Evolution of the Term Chancre and its Relation to the History of Syphilis', *Journal of the History of Medicine* (1949), 407–17.

Keller, Eve, '"That Sublimest Juyce in our Body": Bloodletting and Ideas of the Individual in Early Modern England', *Philological Quarterly* 86:1/2 (2007), 97–123.

Kern Paster, Gail, *The Body Embarrassed: Drama and the Disciplines of Shame in Early Modern England* (Ithaca, NY: Cornell University Press, 1993).

—— 'Melancholy Cats, Lugged Bears, and Early Modern Cosmology: Reading Shakespeare's Psychological Materialism Across the Species Barrier', in Gail Kern Paster, Katherine Rowe and Mary Floyd-Wilson (eds), *Reading the Early Modern Passions: Essays in the Cultural History of Emotion* (Philadelphia: University of Pennsylvania Press, 2004), pp. 113–29.

—— 'The Unbearable Coldness of Female Being: Women's Imperfection and the Humoral Economy', *English Literary Renaissance* 28 (1998), 416–40.

Kerwin, William, *Beyond the Body: The Boundaries of Medicine and English Renaissance Drama* (Massachusetts: University of Massachusetts Press, 2005).

Kitzes, Adam H., *The Politics of Melancholy from Spenser to Milton* (New York: Routledge, 2006).

Klawiter, Maren, *The Biopolitics of Breast Cancer: Changing Cultures of Disease and Activism* (Minneapolis: University of Minnesota Press, 2008).

Kleinmann, Ruth, 'Facing Cancer in the Seventeenth Century. The Last Illness of Anne of Austria, 1644–1666', *Advances in Thanatology* 4 (1978), 37–55.

Kuriyama, Shigehisa, *The Expressiveness of the Body and the Divergence of Greek and Chinese Medicine* (New York; London: Zone Books, 1999).

Kusukawa, Sachiko, 'The Medical Renaissance of the Sixteenth Century: Vesalius, Medical Humanism and Bloodletting', in Peter Elmer (ed.), *The Healing Arts: Health, Disease and Society in Europe, 1500–1800* (Manchester: Manchester University Press, 2004), pp. 58–83.

—— 'Medicine in Western Europe in 1500', in Peter Elmer (ed.), *The Healing Arts: Health, Disease and Society in Europe, 1500–1800* (Manchester: Manchester University Press, 2004), pp. 1–26.

Lacqueur, Thomas, *Making Sex: Body and Gender from the Greeks to Freud* (Cambridge, MA; London: Harvard University Press, 1990).

Lane, Joan, '"The Doctor Scolds Me": The Diaries and Correspondence of Patients in Eighteenth-Century England', in Roy Porter (ed.), *Patients and Practitioners: Lay Perceptions of Medicine in Pre-Industrial Society* (Cambridge: Cambridge University Press, 1985), pp. 205–48.

Langley, Eric, 'Plagued by Kindness: Contagious Sympathy in Shakespearean Drama', *Medical Humanities* 37 (2011), 103–10.

Lanum, Faith, 'Perdita Woman: Sarah Cowper', at *The Perdita Project* (online resource), http://web.warwick.ac.uk/english/perdita, 9 October 2010.

Lee, H.S.J. (ed.), *Dates in Oncology* (Carnforth, UK; Pearl River, NY: Parthenon Publishing Group, 2000).

LeJacq, Seth Stein, 'The Bounds of Domestic Healing: Medical Recipes, Storytelling and Surgery in Early Modern England', *Social History of Medicine* 26:3 (2013), 451–68.

Leong, Elaine, and Sara Pennell, 'Recipe Collections and the Currency of Medical Knowledge in the Early Modern "Medical Marketplace"', in Mark S.R. Jenner and Patrick Wallis (eds), *Medicine and the Market in England and its Colonies, c.1450–c.1850* (Basingstoke; New York: Palgrave Macmillan, 2007), pp. 133–53.

Lerner, Baron H., *The Breast Cancer Wars: Fear, Hope, and the Pursuit of a Cure in Twentieth-Century America* (New York: Oxford University Press, 2001).

Lewison, Edward F., 'Saint Agatha, the Patron Saint of Diseases of the Breast, in Legend and Art', *Bulletin of the History of Medicine* 24 (1950), 409–20.

Lindemann, Mary, *Medicine and Society in Early Modern Europe* (second edition) (Cambridge; New York: Cambridge University Press, 2010).

Löwy, Ilana, *A Woman's Disease: The History of Cervical Cancer* (New York; Oxford: Oxford University Press, 2011).

Lund, Roger, 'Infectious Wit: Metaphor, Atheism and the Plague in Eighteenth-Century London', *Literature and Medicine* 22:1 (2003), 45–64.

MacGregor, Alasdair B., 'The Search for a Chemical Cure for Cancer', *Medical History* 10 (1966), 374–85.

MacInnes, Ian, 'The Politic Worm: Invertebrate Life in the Early Modern English Body', in Jean E. Feerick and Vin Nardizzi (eds), *The Indistinct Human in Renaissance Literature* (Basingstoke: Palgrave Macmillan, 2012), pp. 253–74.

Maclean, Ian, *Logic, Signs and Nature in the Renaissance: The Case of Learned Medicine* (Cambridge: Cambridge University Press, 2002).

Magee, Reginald, 'Saints in Surgery', *Australian and New Zealand Journal of Surgery* 68:8 (1998), 605–10.

Mansfield, Carl M., *Early Breast Cancer: Its History and Results of Treatment*, Series: *Experimental Biology and Medicine* (Basel; New York: Karger, 1976), vol. 5.

Martensen, Robert, 'The Transformation of Eve: Women's Bodies, Medicine and Culture in Early Modern England', in Roy Porter and Mikulas Teich (eds), *Sexual Knowledge, Sexual Science: The History of Attitudes to Sexuality* (Cambridge: Cambridge University Press, 1994), pp. 107–33.

Mazzio, Carla, 'Shakespeare and Science, c.1600', *South Central Review* 26:1–2 (2009), 1–23.

McAllister, Marie E., 'Stories of the Origin of Syphilis in Eighteenth-Century England: Science, Myth, and Prejudice', *Eighteenth-Century Life* 24:1 (2000), 22–44.

McClive, Cathy, 'The Hidden Truths of the Belly: The Uncertainties of Pregnancy in Early Modern Europe', *Social History of Medicine* 15:2 (2002), 209–27.

—— 'Menstrual Knowledge and Medical Practice in Early Modern France, c.1555–1761', in Andrew Shail and Gillian Howie (eds), *Menstruation: A Cultural History* (London: Palgrave Macmillan, 2005), pp. 76–89.

McConchie, R.W., *Lexicography and Physicke: The Record of Sixteenth-Century English Medical Terminology* (New York: Oxford University Press, 1997).

McShane Jones, Angela, 'Revealing Mary', *History Today* (March 2004), 40–46.

McVaugh, Michael, 'Richard Wiseman and the Medical Practitioners of Restoration London', *Journal of the History of Medicine and Allied Sciences* 63:2 (2007), 125–40.

Milburn, Colin, 'Syphilis in Faerie Land: Edmund Spenser and the Syphilography of Elizabethan England', *Criticism* 46:4 (2004), 597–632.

Miles, Margaret R., *A Complex Delight: The Secularization of the Breast 1350–1750* (Berkeley: University of California Press, 2008).

Mortimer, Ian, *The Dying and the Doctors: The Medical Revolution in Seventeenth-Century England* (Woodbridge; Rochester, NY: Royal Historical Society, 2009).

Mukherjee, Siddhartha, *The Emperor of All Maladies: A Biography of Cancer* (London: Scribner, 2010).

Nagy, Doreen Evenden, *Popular Medicine in Seventeenth-Century England* (Bowling Green, OH: Bowling Green State University Popular Press, 1988).

Neely, Carol Thomas, '"Documents in Madness": Reading Madness and Gender in Shakespeare's Tragedies and Early Modern Culture', *Shakespeare Quarterly* 42:3 (1991), 315–38.

Neri, Janice. *The Insect and the Image: Visualizing Nature in Early Modern Europe, 1500–1700* (Minneapolis: University of Minnesota, 2011).

Newton, Hannah, *The Sick Child in Early Modern England* (Oxford: Oxford University Press, 2012).

Ng, Su Fang, *Literature and the Politics of Family in Seventeenth-Century England* (Cambridge: Cambridge University Press, 2007).

Nussbaum, Felicity, *The Brink of All We Hate: English Satires on Women 1660–1750* (Lexington: The University Press of Kentucky, 1984).

—— *Torrid Zones: Maternity, Sexuality, and Empire in Eighteenth-Century English Narratives* (Baltimore, MD; London: The John Hopkins University Press, 1995).

Nutton, Vivian, 'The Seeds of Disease: An Explanation of Contagion and Infection from the Greeks to the Renaissance', *Medical History* 27 (1983), 1–34.

Olson, James S., *Bathsheba's Breast: Women, Cancer and History* (Baltimore, MD: Johns Hopkins University Press, 2002).

Orland, Barbara, '"White Blood and Red Milk": Analogical Reasoning in Medical Practice and Experimental Physiology (1560–1730)', in Manfred Horstmanshoff, Helen King and Claus Zittel (eds), *Blood, Sweat and Tears: The Changing Concepts of Physiology from Antiquity into Early Modern Europe* (Leiden: Brill, 2012), pp. 443–78.

Owens, Sarah E., 'The Cloister as Therapeutic Space: Breast Cancer Narratives in the Early Modern World', *Literature and Medicine* 30:2 (2012), 319–38.

Pack, George T., 'St Peregrine, O.S.M. – The Patron Saint of Cancer Patients', *CA: A Cancer Journal for Clinicians* 17 (1967), 183–4.

Park, Katherine, *Secrets of Women: Gender, Generation and the Origins of Human Dissection* (New York: Zone Books, 2010).

Patterson, James T., *The Dread Disease: Cancer and Modern American Culture* (Cambridge, MA: Harvard University Press, 1987).

Pelling, Margaret, 'Thoroughly Resented? Older Women and the Medical Role in Early Modern London' in Lynette Hunter and Sarah Hutton (eds), *Women, Science and Medicine 1500–1700: Mothers and Sisters of the Royal Society* (Stroud: Sutton Publishing, 1997), pp. 63–88.

Pelling, Margaret (with Frances White), *Medical Conflicts in Early Modern London: Patronage, Physicians, and Irregular Practitioners, 1550–1640* (Oxford: Clarendon Press, 2003).

Phillippy, Patricia, *Women, Death and Literature in Post-Reformation England* (Cambridge: Cambridge University Press, 2002).

Pollard, Tanya, 'Enclosing the Body: Tudor Conceptions of Skin', in Kent Cartwright (ed.), *A Companion to Tudor Literature* (Oxford: Wiley-Blackwell, 2009), pp. 111–23.

Pollock, Linda A., '"Teach her to Live Under Obedience": The Making of Women in the Upper Ranks of Early Modern England', *Continuity and Change* 4:2 (1989), 231–58.

—— *With Faith and Physic: The Life of a Tudor Gentlewoman, Lady Grace Mildmay 1552–1620* (London: Collins and Brown Ltd., 1993).

Pomata, Gianna, 'Menstruating Men: Similarity and Difference of the Sexes in Early Modern Medicine', in Valeria Finucci and Kevin Brownlee (eds), *Generation and Degeneration: Tropes of Reproduction in Literature and History from Antiquity to Early Modern Europe* (Durham, NC; London: Duke University Press, 2001), pp. 108–52.

Porter, Roy, *The Greatest Benefit to Mankind: A Medical History of Humanity from Antiquity to the Present* (London: Harper Collins, 1999).

—— *Patients and Practitioners: Lay Perceptions of Medicine in Pre-Industrial Society* (Cambridge: Cambridge University Press, 1985).

—— (ed.), *The Popularization of Medicine 1650–1850* (London: Routledge, 1992).

Porter, Roy, and Dorothy Porter, *In Sickness and in Health: The British Experience 1650–1850* (London: Fourth Estate, 1988).

Pouchelle, Marie-Christine, *The Body and Surgery in the Middle Ages* (trans. Rosemary Morris), (Cambridge: Polity Press, 1990).

Power, D'A., 'Needham, Walter (*bap.* 1632, *d.* 1691)', rev. Patrick Wallis, *Oxford Dictionary of National Biography*, Oxford University Press, 2004; online edition, May 2006, http://www.oxforddnb.com/view/article/19849, 1 July 2013.

Purnis, Jan, 'The Stomach and Early Modern Emotion', *University of Toronto Quarterly* 79:2 (2010), 801–18.

Qualtiere, Louis F., and William W.E. Slights, 'Contagion and Blame in Early Modern England: The Case of the French Pox', *Literature and Medicine* 22:1 (2003), 1–24.

Radden, Jennifer, *The Nature of Melancholy: From Aristotle to Kristeva* (Oxford: Oxford University Press, 2002).

Rather, Lelland J., *The Genesis of Cancer: A Study in the History of Ideas* (Baltimore, MD: Johns Hopkins University Press, 1978).

Reedy, Jeremiah, 'Galen on Cancer and Related Diseases', *Clio Medica* 10 (1975), 227–38.

Rockwood, Camilla (ed.), *Brewer's Dictionary of Phrase & Fable* (eighteenth edition) (London: Chambers Harrap Publishers Ltd., 2010).

Rosenberg, Charles E., 'Disease in History: Frames and Framers', *The Milbank Quarterly* 67 (Supplement 1: *Framing Disease: The Creation and Negotiation of Explanatory Schemes*) (1989), 1–15.

Sakorafas, George H., and Michael Safioleas, 'Breast Cancer Surgery: An Historical Narrative. Part I. From Prehistoric Times to Renaissance', *European Journal of Cancer Care* 18:6 (2009), 530–44.

Salmon, Marylynn, 'The Cultural Significance of Breastfeeding and Infant Care in Early Modern England and America', *Journal of Social History* 28:2 (1994), 247–69.

Sawday, Jonathan, *The Body Emblazoned: Dissection and the Human Body in Renaissance Culture* (London: Routledge, 1995).

Scarry, Elaine, *The Body in Pain: The Making and Unmaking of the World* (New York; Oxford: Oxford University Press, 1985).

Schiebinger, Londa, 'Skeletons in the Closet: The First Illustrations of the Female Skeleton in Eighteenth-Century Anatomy', *Representations* 14, *The Making of the Modern Body: Sexuality and Society in the Nineteenth Century* (1986), 42–82.

Schliener, Winfried, *Medical Ethics in the Renaissance* (Washington, DC: Georgetown University Press, 1995).

Schoenfeldt, Michael, 'Aesthetics and Anaesthetics: The Art of Pain Management in Early Modern England', in Jan Frans Van Dijkhuizen and Karl A.E. Enenkel (eds), *The Sense of Suffering; Constructions of Physical Pain in Early Modern Culture* (Leiden; Boston: Brill, 2009), pp. 19–38.

—— *Bodies and Selves in Early Modern England: Physiology and Inwardness in Spenser, Shakespeare, Herbert, and Milton* (Cambridge: Cambridge University Press, 1999).

Semonin, Paul, 'Monsters in the Marketplace: The Exhibition of Human Oddities in Early Modern England', in Rosemarie Garland Thomson (ed.), *Freakery: Cultural Spectacles of the Extraordinary Body* (New York; London: New York University Press, 1996), pp. 69–81.

Shepard, Alexandra, *Meanings of Manhood in Early Modern England (Oxford Studies in Social History)* (Oxford: Oxford University Press, 2003).

Shimkin, Michael B., *Contrary to Nature: Being an Illustrated Commentary on Some Persons and Events of Historical Importance in the Development of Knowledge Concerning Cancer* (Washington: 1977).

Shorter, Edward, *A History of Women's Bodies* (Harmondsworth: Pelican, 1984).

Siena, Kevin P., 'Pollution, Promiscuity, and the Pox: English Venereology and the Early Modern Medical Discourse on Social and Sexual Danger', *Journal of the History of Sexuality* 8:4 (1998), 553–74.

—— (ed. and Introduction), *Sins of the Flesh: Responding to Sexual Disease in Early Modern Europe* (Toronto: Centre for Reformation and Renaissance Studies, 2005).

—— *Venereal Disease, Hospitals and the Urban Poor: London's 'Foul Wards', 1600–1800* (Rochester, NY: University of Rochester Press, 2004).

Silverman, Lisa, *Tortured Subjects: Pain, Truth and the Body in Early Modern France* (Chicago: The University of Chicago Press, 2001).

Siraisi, Nancy G., *Medieval and Early Renaissance Medicine: An Introduction to Knowledge and Practice* (Chicago: University of Chicago Press, 1990).

Slack, Paul, 'Mirrors of Health and Treasures of Poor Men: The Uses of the Vernacular Medical Literature of Tudor England', in Charles Webster (ed.), *Health, Medicine and Mortality in the Sixteenth Century* (Cambridge: Cambridge University Press, 1979), pp. 237–75.

Smyth, Adam, 'Almanacs, Annotators, and Life-Writing in Early Modern England', *English Literary Renaissance* 38:2 (2008), 200–44.

Sontag, Susan, *Illness as Metaphor and AIDS and Its Metaphors* [London: Penguin, 1991 (Illness as Metaphor first published 1977, AIDS and Its Metaphors first published 1988)].

Stallybrass, Peter, 'Patriarchal Territories: The Body Enclosed', in Margaret W. Ferguson, Maureen Quilligan and Nancy J. Vickers (eds), *Rewriting the Renaissance: The Discourses of Sexual Difference in Early Modern Europe* (Chicago; London: University of Chicago Press, 1986), pp. 123–42.

Stephenson Payne, Lynda Ellen, *With Words and Knives: Learning Medical Dispassion in Early Modern England* (Aldershot: Ashgate, 2007).

Stolberg, Michael, 'Active Euthanasia in Pre-Modern Society, 1500–1800: Learned Debates and Popular Practices', *Social History of Medicine* 20:2 (2007), 205–21.

—— *Experiencing Illness and the Sick Body in Early Modern Europe* (Basingstoke: Palgrave Macmillan, 2011).

—— 'Menstruation and Sexual Difference in Early Modern Medicine', in Andrew Shail and Gillian Howie (eds), *Menstruation: A Cultural History* (London: Palgrave Macmillan, 2005), pp. 90–101.

—— 'A Woman Down to Her Bones: The Anatomy of Sexual Difference in the Sixteenth and Early Seventeenth Centuries', *Isis* 94:2 (2003), 274–99.

—— 'A Woman's Hell? Medical Perceptions of Menopause in Pre-Industrial Europe', *Bulletin of the History of Medicine* 73 (1999), 408–28.

Stone, Lawrence, *Broken Lives: Separation and Divorce in England 1660–1857* (Oxford: Oxford University Press, 1993).

—— *The Family, Sex and Marriage in England, 1500–1800* (New York: Harper & Row, 1977).

Sugg, Richard. *Murder after Death: Literature and Anatomy in Early Modern England* (Ithaca, NY: Cornell University Press, 2007).

Taylor, Charles, *Sources of the Self: the Making of the Modern Identity* (Cambridge, MA: Cambridge University Press, 1989).

Thirsk, Joan, *Food in Early Modern England: Phases, Fads, Fashions, 1500–1760* (London: Continuum, 2009).

Thomas, Keith, *Man and the Natural World: Changing Attitudes in England, 1500–1800* (London: Penguin, 1983).

Thompson, Pauline, 'The Disease That We Call Cancer', in S. Campbell, B. Hall and D. Klausner (eds), *Health, Disease and Healing in Medieval Culture* (Basingstoke: Palgrave Macmillan, 1992), pp. 1–11.

Totaro, Rebecca, *Suffering in Paradise: The Bubonic Plague in Literature from More to Milton* (Pittsburgh, PA: Duquesne University Press, 2005).

Toulalan, Sarah, *Imagining Sex: Pornography and Bodies in Seventeenth-Century England* (Oxford: Oxford University Press, 2007).

—— '"To[o] Much Eating Stifles the Child": Fat Bodies and Reproduction in Early Modern England', *Historical Research* 87:235 (2014), 67–93.

Van Dijkhuizen, Jan Frans and Karl A.E. Enenkel, 'Introduction: Constructions of Physical Pain in Early Modern Culture', in Jan Frans Van Dijkhuizen and Karl A.E. Enenkel (eds), *The Sense of Suffering; Constructions of Physical Pain in Early Modern Culture* (Leiden; Boston: Brill, 2009), pp. 1–18.

Van Helden, Albert, 'The Invention of the Telescope', *Transactions of the American Philosophical Society*, New Series 67:4 (1977), 1–67.

Walker, Garthine, *Crime, Gender, and Social Order in Early Modern England* (New York: Cambridge University Press, 2003).

Watkins, Calvert, *How to Kill a Dragon: Aspects of Indo-European Poetics* (Oxford: Oxford University Press, 1995).

Wear, Andrew, *Knowledge and Practice in English Medicine, 1550–1680* (Cambridge: Cambridge University Press, 2000).

—— 'Medical Ethics in Early Modern England', in Andrew Wear, Johanna Geyer-Kordesch and Roger French (eds), *Doctors and Ethics: The Earlier Historical Setting of Professional Ethics* (Amsterdam: Rodopi, 1993), pp. 98–130.

—— 'Puritan Perceptions of Illness in Seventeenth Century England', in Roy Porter (ed.), *Patients and Practitioners: Lay Perceptions of Medicine in Pre-Industrial Society* (Cambridge: Cambridge University Press, 1985), pp. 54–99.

Weber, A.S., 'Women's Early Modern Medical Almanacs in Historical Context', *English Literary Renaissance* 33:3 (2003), 358–402.

Webster, Charles, 'Paracelsus: Medicine as Popular Protest', in Peter Ole Grell and Andrew Cunningham (eds), *Medicine and the Reformation* (Abingdon, Oxon: Routledge, 1993), pp. 57–77.

Weisser, Olivia, '(Roy Porter Student Prize Essay) Boils, Pushes and Wheals: Reading Bumps on the Body in Early Modern England', *Social History of Medicine* 22:2 (2009), 321–39.

Williams, Gordon, *Shakespeare's Sexual Language: A Glossary (Volume III: Q–Z)* (London; Atlantic Highlands, New Jersey: The Athlone Press, 1994).

Wilson, Catherine, *The Invisible World: Early Modern Philosophy and the Invention of the Microscope* (Princeton, NJ: Princeton University Press, 1997).

Wilson, Miranda. 'Watching Flesh: Poison and the Fantasy of Temporal Control in Renaissance England', *Renaissance Studies* 27:1 (2013), 97–113.

Wilson, Philip K., *Surgery, Skin and Syphilis: Daniel Turner's London (1667–1741)* (Amsterdam; Atlanta: Rodopi, 1999).

Wishart, Adam, *One in Three: The Discovery of Cancer and the Search for a Cure* (London: Profile, 2006).

Wood, Andy, *Riot, Rebellion and Popular Politics in Early Modern England* (Basingstoke: Palgrave, 2002).

Wright, Jonathan, 'The World's Worst Worm: Conscience and Conformity during the English Reformation', *The Sixteenth Century Journal* 30:1 (1999), 113–33.

Wright, Thomas, *Circulation: William Harvey's Revolutionary Idea* (London: Vintage, 2013).

Wyman, A.L., 'The Surgeoness: The Female Practitioner of Surgery 1400–1800', *Medical History* 28 (1984), 22–41.

Yalom, Marilyn, *A History of the Breast* (London: Pandora, 1998).

Index